Science and the Navy

Science and the Navy:

THE HISTORY OF THE OFFICE
OF NAVAL RESEARCH

Harvey M. Sapolsky

PRINCETON UNIVERSITY PRESS

PRINCETON, NEW JERSEY

Copyright © 1990 by Princeton University Press
Published by Princeton University Press, 41 William Street,
Princeton, New Jersey 08540
In the United Kingdom: Princeton University Press, Oxford

Library of Congress Cataloging-in-Publication Data

Sapolsky, Harvey M.
Science and the Navy : the history of the Office of Naval Research / Harvey M.
Sapolsky.
p. cm.
1. United States. Office of Naval Research—History. 2. Science and state—United
States—History. 3. Federal aid to research—United States—History. I. Title.
V394.A7S37   1990   359'.07'0973—dc20   89-48144

ISBN 0-691-07847-5 (alk. paper)

The preparation of this work was partially supported by the Department
of the Navy under Contract Number: N000 14-16-A-0204-0046; however,
the conclusions and opinions stated in this work are those of
the author, and do not necessarily represent or reflect the views
of the Department of the Navy.

To Karen

Contents

List of Figures and Tables

Preface

THERE ARE many ways to be blessed—good health, good looks, and a good marriage being among the most frequently cited. I believe that possessing patient benefactors should be included on that treasured list. The Office of Naval Research and the Alfred P. Sloan Foundation, the organizations which shared the burden of financing this research project, are certainly patient because this work began more than a decade ago and continued long after its originally promised conclusion. For their help and understanding, I am grateful.

Of course, I had excuses. Some of my most creative writing occurred as I sought to explain the growing gap between due dates and accomplishment. The truth was that I became diverted by a developing interest in health policy, and one project led to another. Although I never quite solved all of the dilemmas of this field, I did eventually satisfy my quest to learn as much as I could about the forces that influence society's health policy choices.

It is a special privilege of an academic career to be able to indulge one's curiosity. Those who work in more bureaucratic settings are denied the opportunity to follow interest rather than schedule. Not surprisingly, they often resent academics who can do this, thinking them lazy and unreliable. Although these faults may in fact apply to me, I did not forget an unfulfilled obligation. The guilt of not completing entirely a promised task soured the joy of finding new ones. I had to return to this work even though the formal requirement to do so had been waived.

As I explore again the issues involved in the early history of the Office of Naval Research, I am struck by how relevant they remain. America has just experienced its fourth cycle of growth in defense spending since the Second World War. Once more, defense support of basic research has correspondingly increased. And again, perhaps predictably, strains developed in the relationship between university-based scientists and the military. Sorting through what happened in the past gives perspective on what is happening in the present.

Although without the bitterness of the Vietnam era, the issues of the late 1960s are again with us. Science is wealthier than it has been in recent years, in part because scientific progress is described again as enhancing national security. The wealth is welcome within the universities, but the link to national security is not. Science administrators in government, for reasons that I explain in this volume, exaggerate the immediacy and strength of the tie between science and improved military capabilities. The agencies that support science operate in a world of planning documents where the exaggerations are codified. Opponents of American foreign and defense policies—of which there are many in the nation's universities—use the language of the documents as proof that scientific values are being perverted by military support of science. And at the highest levels of government, those who are committed to the policies use the very same language as justification, both to restrict access to scientific information and to deny funds to likely critics.

With this study of the first defense agency to support academic science after the Second World War, I hope to increase the understanding of how science and the military have interacted in this unusual era in the American experience. The history of the Office of Naval Research provides an important opportunity to review the effects of defense support on both military preparedness and academic freedom. Today's debates repeat to a large extent the debates that greeted the military's new role in financing university science. The difference is that too few who participate in today's debate know the history.

I must acknowledge the assistance that I received. The late Peter King, Chief Scientist of the Office of Naval Research when I began my work, was immensely helpful in providing access and guidance. As a civilian scientist in the Navy in the late 1940s, Dr. King detected the Soviet Union's first test explosion of a nuclear weapon. With his career in defense research, he sought to ensure that such weapons would never be used. Dr. Sidney G. Reed Jr., who also had a distinguished career in research administration in both the Office of Naval Research and the National Science Foundation, served for several years as my project monitor. He contributed greatly to this study by sharing his deep insight into the administration of research in military organizations and outside of them. Dr. Bruce S. Old, who played a key role in the founding and development of the Office, was most helpful.

During the course of the work, I interviewed about 250 scientists, military officers, and government officials. Most agreed to cooperate, and several even opened their personal files to inspection. I am grateful for their generous assistance, without which I could not have completed this project, but I will not identify them here, because the individuals interviewed for this work were promised anonymity in any resulting publication. (The citations of interviews in the notes refer to my records and coding system. These records will be preserved and made available for inspection by interested scholars.)

Early in the study, my friend and former student, Lawrence Mc-Cray, provided valuable research assistance. He is now on the staff of the National Research Council at the National Academy of Sciences. Another friend and former student, Sanford L. Weiner, was also involved in the initial phases of the study. One of Sandy's publications, based in part on this work, is cited extensively in the text. More recently, I was aided by the assistance of Jamie Aisenberg, a physician-in-training who has catholic interests; Shannon Kile, a doctoral candidate in political science at the Massachusetts Institute of Technology; and Hilary Coller, a Harvard undergraduate. Judith Spitzer once again was at the keyboard. Her tolerance for poor handwriting, atrocious spelling, and lame attempts at humor is legendary.

No one in science policy has combined scholarship and firsthand experience as effectively as Eugene Skolnikoff, my colleague in political science at the Massachusetts Institute of Technology. In this study, as with so much of my work, I am in his debt. Gene helped arrange the project and provided constructive advice throughout its long gestation. It is my good fortune that his sound judgment and generous encouragement were close at hand.

Parts of this work were presented at conferences and seminars. I thank Nathan Reingold for including my summary essay in the bicentennial volume on American science that he put together for the Smithsonian Institution Press; William Leslie of Johns Hopkins University for inviting my participation in a symposium on science and the military that he organized; Daniel M. Masterson of the History Department at the U.S. Naval Academy for including me in the Sixth Symposium on Naval History that he arranged; Paul Hoch of the University of Warwick and Everett Mendelsohn of Harvard University for asking me to join their panel on science and the military at the Manchester meetings of the British and U.S. History of Science Societies. All of them improved upon the work. As is his habit, Dean C. Allard of the Naval Historical Center was kind to a political scientist posing as a historian. And David Allison, an expert on the development of radar and other aspects of naval technology, was most considerate in providing comments and suggestions.

None of the above, including the project's benefactors, should be held responsible for the interpretations of the Office of Naval Research's experience that follow. I too might wish that status, but know it is not possible. And then again, there is a certain pleasure in believing that you understand history even when others believe differently.

Belmont, Massachusetts

Abbreviations

AEC	Atomic Energy Commission
AFOSR	Air Force Office of Scientific Research
ARDC	Air Research and Development Command
ARPA	Advanced Research Projects Agency
BOB	Bureau of the Budget
CNA	Center for Naval Analyses
CNO	Chief of Naval Operations
DARPA	Defense Advanced Research Projects Agency
DOD	Department of Defense
FY	Fiscal Year
INS	Institute for Naval Studies
M.I.T.	Massachusetts Institute of Technology
NACA	National Advisory Committee for Aeronautics
NAS	National Academy of Sciences
NASA	National Aeronautics and Space Administration
NAVWAG	Naval Weapons Analysis Group
NDRC	National Defense Research Committee
NIH	National Institutes of Health
NRAC	Naval Research Advisory Committee
NRL	Naval Research Laboratory
NSF	National Science Foundation
OEG	Operations Evaluation Group
ONR	Office of Naval Research
OPNAV	Chief of Naval Operations' Staff
OP-06	Deputy Chief of Naval Operations for Special Weapons
ORI	Office of Research and Inventions
OSRD	Office of Scientific Research and Development
RBNS	Research Board for National Security
R&D	Research and Development
RDB	Research and Development Board, first Joint Research and Development Board
SAB	Science Advisory Board of the Air Force
TRACES	Technology in Retrospect and Critical Events in Science

Science and the Navy

CHAPTER ONE

Introduction

AMERICAN indifference to basic research, a common lament among historians and scientists alike, can be exaggerated.[1] America supports an impressive array of scientific institutions and maintains an army of researchers at an unparalleled cost. American scientists have won a disproportionate share of the world's scientific prizes. It is their work, much more than that of the British, that has made English the language of science.

To be sure, the federal government until recently has not played a dominating role in developing the nation's scientific capabilities.[2] It was state governments competing for economic advantage within the union or wealthy individuals hoping for permanent remembrance that gave American science most of its resources up to the Second World War.[3] When the federal government did provide assistance, it often did so indirectly as in the case of the Morrill Act which with land grants encouraged the states to establish institutions of higher education.

Invariably, the beneficence provided science was made on visions of expanded commerce and human betterment. These expectations were, of course, encouraged by the scientists themselves who both believed in them and knew their power to entice potential patrons.[4] The accounting was hardly exact. Scientists managed to pursue their interest in expanding knowledge with whatever rationale that was expedient.[5]

[1] Nathan Reingold, "American Indifference to Basic Research," in George H. Daniels, ed., *Nineteenth Century American Science: A Reappraisal* (Evanston, Ill.: Northwestern University Press, 1972). See also George H. Daniels, "The Pure-Science Ideal and Democratic Culture," *Science* 156 (30 June 1967): 1699–1705.

[2] A. Hunter Dupree, *Science in the Federal Government* (Cambridge, Mass.: Harvard University Press, 1957). For a brief overview of U.S. science policy, see Jeffrey K. Stine, *A History of Science Policy in the United States 1940–1985*, a report prepared for the Task Force on Science Policy, Committee on Science and Technology, House of Representatives, 99th Congress, 2d Session, September 1986.

[3] Howard S. Miller, *Dollars for Research: Science and Its Patrons in Nineteenth Century America* (Seattle, Wash.: University of Washington Press, 1970).

[4] Ibid.; Michael D. Reagan, *Science and the Federal Patron* (New York: Oxford University Press, 1969).

[5] William Culp Darrah, *Powell of the Colorado* (Princeton, N.J.: Princeton University

One rationale, however, that was rarely invoked prior to the Second World War was that science could enhance national security. Geography, not military force, was our defense. Two oceans protected us and gave much freedom from serious foreign threats. Only during brief periods of conflict was science mobilized, and then not extensively or with great effect either upon science itself or the military outcome.

The Second World War changed all of this. It was the war that made the United States a major world power, complete with the need for large standing forces. It was also in Churchill's phrase "the Wizard War," the conflict in which the world discovered that effective military action was significantly dependent upon scientific and technological advances. As a consequence, the federal government took on new responsibilities in our society, not the least of which was the patronage of research and development activities with the intent of making American forces the strongest on earth.

The scale of the effort has been enormous.[6] Hundreds of billions of dollars have been invested in military research, most of which was directed toward the development of specific weapon systems and conducted by industrial contractors. But no element of the research enterprise, including basic research, has been untouched by this largess. For better or worse, American science became both richer and more closely linked to the nation's defense in the years since the Second World War than it had ever been.

Here I examine the origins and experience of the agency most responsible for forging that link, the Office of Naval Research (ONR). As the first (and for several years the only) federal agency to support a broad range of scientific work in the universities, ONR helped design the structure of government-science relations that stands today. The administrative mechanisms and political justifications it initially selected for the support of basic research are still with us. So too is the skepticism that greeted its proclaimed intent to preserve the freedom of scientific inquiry while financing scientific investigations that would serve the operational needs of the Navy.

Three questions have guided my study. First, I wondered why it was the Navy that took the lead within the government in supporting

---

Press, 1951); Walter B. Hendrickson, "Nineteenth-Century Geological Survey: Early Government Support of Science," *Isis* 52 (1961): 357–71; Gerald D. Nash, "The Conflict Between Pure and Applied Research in Nineteenth Century Public Policy," *Isis* 54 (1963): 217–28.

[6] Harvey M. Sapolsky, "Science, Technology, and Military Policy," in I. Speigel-Rosing and D. de Solla Price, eds., *Science Policy Studies in Perspective* (Beverly Hills, Calif.: Sage, 1977).

basic research. The Navy may seem the obvious candidate as it has been the most technologically oriented of the two original military departments and at least a strong rival to the Air Force for the current title. Survival at sea pits man against nature more than it does man against man. It is not surprising that naval officers would do pioneering work in navigation, meteorology, and engineering and that the Navy would build enduring scientific institutions such as the Naval Observatory and the Naval Research Laboratory. Naval officers helped found the National Academy of Sciences; and one, a graduate of Annapolis, A. A. Michelson, won America's first Nobel Prize for his work in physics. Moreover, the construction and maintenance of ships requires the preservation within the Navy of engineering skills firmly based on scientific disciplines. So too does the Navy's interest in communications, aircraft, and missiles. But nearly all of the Navy involvement in research and technical training prior to the Second World War was confined to the Navy's shore establishment which had built an elaborate network of laboratories and test facilities. Because of the strength of its own capabilities, it was precisely an organization that would be unlikely to turn to others for advice and assistance on scientific matters.

Second, I was curious how the Navy's largess, once offered, was received. The divisions in American higher education between public and private institutions, the religious and the secular, and the elite and the commonplace are sharp, if not always acknowledged. I wondered how these divisions had affected the distribution of research allocations. In addition, scientists and universities were anxious at the end of a long war to have the military's security restrictions removed from campuses. There was even concern in the mid-1940s that the universities would become dependent upon federal funds. Nevertheless, the funds were accepted, including those provided by the military. In the late 1960s and early 1970s the universities were in turmoil over the Vietnam War, one consequence being that there was growing opposition to the military's sponsorship of campus-based research, but little analysis of the reasons why it existed. It seemed important to gain an understanding of the policy decisions that had been made by both the military and the universities in building this troubled relationship.

Third, there is the question of the actual contribution of basic research to national security. From my previous work in science policy, I knew that there were persistent doubts within the military about the benefits of supporting basic research. The continual citation by scientists of particular weapons achievements such as the atomic bomb and radar were apparently unpersuasive, both in terms of their historical accuracy and as a guide for research management. Even more problematic for the military was the participation of scientists in the formation of high-level defense policy. Some scientists, especially those who had a role in the major weapons projects of the Second World War, had come to serve as senior governmental advisers and were often the strongest advocates for increased military allocations for basic research. A review of the ONR experience requires an analysis of the military's use of basic research results and the effect scientists can have on policy because the agency chose to be the advocate of both basic research and the academic scientist within the Navy.

A risk in any discussion of research is semantic confusion due to the many possible modifiers of the word "research" that can appear. Lord Rothschild, in an amusingly written but serious-minded essay, notes that he has collected, in official reports, forty-five varieties ranging from absolutely pure research to basic applied research.[7] Like him, I believe the two main categories are: research carried out by the investigator solely to increase knowledge, and research that is conducted with a potential application in view. Although we agree that the latter category should be called applied research, his preference for the former is pure research, while mine is basic research. Perhaps it is the American in me that causes this deviation. The hope is that the categories are used consistently and that most obscuring variations can be dispensed with (e.g., strategic applied research or applied strategic research). This promise of Rothschildian clarity notwithstanding, I cannot avoid, in what follows, the occasional use of such bureaucratically favored synonyms as fundamental research (for basic) and mission-oriented research (for applied). I trust the good Lord will forgive me. Also I am aware, as we all should be, that there is much grayness between what we call science and what we call technology. At the extremes they are quite different, but they merge easily in the middle.

Data for this study was gathered in a number of ways, none of which represent breakthroughs in social science methodology. Primarily, I relied on unstructured interviews with knowledgeable individuals. Included were present and former staff members of the Office of Naval Research, scientists, naval officers, officials of other federal agencies, and university administrators. About 250 individuals were interviewed in sessions ranging from a half hour to several days (for particularly valued sources). In addition, I had access to the agency's records and several private collections of papers and letters. Official

[7] Lord Rothschild, "Forty-Five Varieties of Research (and Development)," *Nature* 239 (13 October 1972): 373–78.

reports are cited frequently, if only to provide date tags to important events.[8]

My account of ONR's history stresses organizational achievement rather than individual achievements. The triumphs of scientists working under ONR's auspices are relevant, but only to the extent to which they were recognized and used to influence the Office's fate. Some, unknown to me as well as to ONR, may eventually be considered great moments in the history of science even though they played no part in the struggle for budgets and jurisdiction that is central here. Their description must await a different effort which I am sure will be done.

Chapter 2 describes the origin of ONR. Of necessity there is a discussion of the relations between scientists and the military during the Second World War. As is the case for most organizations, there is a mythology surrounding ONR's founding. The reality of that important occurrence is somewhat in variance with the traditional story, as useful as the story may be for various constituencies.

For a few years, in the late 1940s, ONR functioned as the federal government's only general science agency. Chapter 3 examines that period and ONR's role in pioneering the government's relationship with universities. There was an opportunity to consolidate the government's support of science in a single agency when the National Science Foundation was established, but the expansionistic urges were still flowing freely within ONR and it sought to find an independent niche.

The Navy discovered ONR just before the beginning of the Korean War. This was a difficult time for the agency because it had to justify its existence to those whom it thought it had been serving well. The issue of relevance as a pressing organizational problem is introduced in chapter 4. It is an issue that remains throughout the rest of the manuscript.

The research successes and failures of ONR are examined in chapter 5. As the relationship between research and organizational mission is a common one in government, the experience of other agencies is also explored. ONR's unique contribution to the management of research, rarely appreciated within the Navy, was its ability to attract program specialists who did what no management system has been able to do—bridge the large gap that exists between the worlds of the Navy and of academic science. However, the overwhelming desire within

---

[8] Applications for access to official documents, unless otherwise indicated, should be made to the Center for Naval History. The papers of Vannevar Bush are located at M.I.T. and the Library of Congress. Those of Alan Waterman are in the Library of Congress and those of Vice Adm. Harold G. Bowen are at the Seeley G. Mudd Manuscript Library at Princeton University.

the Department of Defense is for a research management system that can be minutely specified and whose performance can be measured.

The scientist's role as high-level policy adviser has intrigued many observers, their conclusion usually being that scientists have had great influence in government decision making. ONR often acted as sponsor for scientists advising the Navy. Chapter 6 examines the advisory experience of these scientists, noting the limited achievements and more common frustrations. The intense institutional loyalties generated within the officer corps in fact made it difficult for scientists to obtain the influence they desired and that others attributed to them.

The Office of Naval Research no longer commands a central role in the formulation of the federal government's science policy. Its influence was diminished by events well beyond its control. To the relief of many scientists, academic research is now much less of a Navy or defense department sponsored enterprise than it was just two decades ago. Chapter 7 places in perspective ONR's transition from the hub to the periphery of policy making. It also assesses the consequences for science of this change. Most scientists may not yet appreciate how much they have lost because the support of science has been civilianized.

However, there is a generation of scientists, the generation trained immediately after the Second World War, that cherishes the memory of a government agency that knew the value of science to society and sought to promote its well being. The Navy's amazingly good sense, demonstrated in the founding of ONR, they believe, ought to continue to be the guide for policy. But naval officers, even those involved in the early years of ONR, remain puzzled by the Navy's nurturing of basic science. As one pointedly asked, "Why did Santa Claus wear a blue suit?"[9]

In life, achievements are more frequently unintended than intended. Those who create organizations do so with particular agendas in mind. If it is a government agency that they create, their control over its future is rarely more than fleeting. ONR was established for one purpose, yet served several. Neither its founders nor their successors quite had their way despite attempts to plan carefully. And yet, notwithstanding its inception as a bureaucratic accident, ONR deserves to be remembered as America's most effective patron of science.

[9] Interview.

The Origins of the Office of Naval Research

NEITHER the U.S. Navy nor any of the other armed services sought the task of supporting science at the end of the Second World War. To be sure, senior military and naval officers were extremely impressed with the rapid progress that had been achieved in weapons during the war. But if on occasion they praised the contribution of science to the war effort, they usually meant the contribution of technology, committing the common verbal error of confusing science and technology, which scientists are not always moved to correct.[1] The task of supporting science, perhaps appropriately, held no special interest for them. The fact that the Office of Naval Research filled a void in the financing of science at the end of the war was quite unintentional; it had been created for a different purpose.

The concept of the Office of Naval Research originated with a group of young reserve officers who served during the war in the Office of the Coordinator of Research and Development, an organization that linked the civilian directed Office of Scientific Research and Development (OSRD) with the Navy's materiel bureaus. Known as the "Bird Dogs" because of their job of ferreting out problems in interorganizational relationships, these young officers spent much of their spare time planning the structure of the Navy's postwar research effort. In their view, the Navy needed a permanent office to coordinate the research programs of its bureaus and to work with civilian scientists in the development of weapons. As early as November 1943, they had evolved a design for a central research office for the Navy which was nearly identical to the design adopted for the Office of Naval Research when it was established at the end of the war.[2] The Bird Dogs sought

---

[1] See, for example, Fleet Adm. Ernest J. King, *U.S. Navy at War 1941–1945: Official Reports of the Secretary of the Navy* (Washington, D.C.: Department of the Navy, 1946), issued 8 December 1945, and "Navy Science Link Stressed as Vital," *New York Times* 8 December 1945, 30. For a contrary view arguing that the military were won over to science, see Alfred Jones, "The National Science Foundation," *Scientific American*, June 1948, 9.

[2] The Bird Dogs (Bruce S. Old), "The Evolution of the Office of Naval Research," *Physics Today*, August 1961, 32. Also Bruce S. Old, "Return on Investment in Basic Research—Exploring a Methodology," *The Bridge*, Spring 1982, 20–26 and John Walsh, "Office of Naval Research Marks 40th Anniversary," *Science* 234 (21 November 1986): 932–33.

to advance their proposal by submitting it in the fall of 1944 to the Secretary of the Navy who was then beginning to consider demobilization plans.[3] Knowing President Franklin D. Roosevelt's personal interest in the organization of the Navy (he was once its Assistant Secretary), they had scheduled a presentation for him on their proposal, but it never took place because of his death.[4] It was not, however, the Bird Dogs' efforts, as bold as they were, that gave the Navy an Office of Naval Research, nor was it their plan for coordination of research that the establishment of the office was intended to implement.

The founder of ONR was Vice Adm. Harold G. Bowen, whose intent was to gain an organizational base from which he could promote the development of nuclear propulsion for naval vessels.[5] A pioneer in the development of high pressure steam power plants for warships, Admiral Bowen believed that nuclear power would be the next major innovation in ship propulsion systems and he wanted to be its pioneer as well. As Director of the Naval Research Laboratory at the start of the Second World War, he was one of the few naval officers who knew about the plan to develop the atomic bomb and the potential that existed for the use of nuclear power in ships. Admiral Bowen's brusque assertion of the Navy prerogatives in weapon research, however, antagonized Vannevar Bush and the other prominent scientists who were attempting to organize a civilian directed weapon development program, and in 1941 he was forced out of his senior post in naval research. Bowen's clash with civilian scientists was in large measure the cause of the Navy being excluded from participation in the Manhattan Project, the effort to develop the atomic bomb.[6] At the end of the war, Admiral Bowen sought again a central role in the Navy's research effort, but this time as the champion of the civilian scientists, by being the director of an agency apparently established to serve their interests. Eventually he lost the struggle for internal Navy jurisdiction for the development of nuclear propulsion systems to a group in the Bureau of Ships led by Captain, later Admiral, Hyman G. Rick-

[3] Memorandum from the Coordinator of Research and Development to the Secretary of the Navy. Subj.: Organization of Research in the Navy Department, 11 October 1944, Draft Memorandum from Lt. Comdr. R. A. Krause, USNR; Lt. Comdr. B. S. Old, USNR; and Lt. J. T. Burwell, Jr., USNR to the Secretary of the Navy via the Coordinator of Research and Development. Subj.: Suggestion for Post-War Organization of Naval Research, 23 September 1944.

[4] The Bird Dogs, "The Evolution of the Office of Naval Research," 35.

[5] See Richard G. Hewlett and Francis Duncan, *Nuclear Navy 1946–1952* (Chicago: University of Chicago Press, 1974), 25.

[6] Vannevar Bush, *Pieces of the Action* (New York: William Morrow, 1970), 104–6. See also n. 43 below.

over and finally had to give up his dream. The legacy Bowen left the Navy was the Office of Naval Research.

Civilian scientists had to overcome the resistance of officers in the Army as well as the Navy in order to gain an independent role in the development of weapons during the Second World War. Having sparked the development of the atomic bomb, the proximity fuse, radar, and dozens of other weapon innovations, these scientists at the end of the war were proud of their contribution and convinced that the military would stifle further innovations unless their independent role in research was permanently guaranteed. A number of organizational arrangements to protect the autonomy of civilian scientists were proposed, but encountering further resistance from the military and important, though unanticipated, constitutional questions, each failed early adoption. Ironically, the only government agency of any significant size available at the war's end to support research of civilian scientists was the one Admiral Bowen founded, the Office of Naval Research.

THE NAVY'S PRE-WAR RELATIONSHIP WITH SCIENTISTS

The Navy had always been the branch of the American military that was the most accommodating to scientists. It had sponsored a greater number of scientific expeditions than the Army during the nineteenth century. Its officers historically were more technically trained than were those of the Army. Its laboratory employees tended to be more accomplished as scientists within government. It, rather than the Army, had established civilian consulting boards in both the Civil War and the First World War. And yet, on the eve of the Second World War, the Navy's relationships with civilian scientists were the most strained of the armed services.

Part of the problem that developed in the relations between civilian scientists and the Navy is attributable to a clash of personalities. Admiral Bowen, as a result of an appointment as Technical Aide to the Secretary of the Navy, was responsible for the Navy's liaison with scientists between the autumn of 1939 and the summer of 1941. Having then just completed a tour of duty as Chief Engineer of the Navy, he was confident of his own technical judgments and those of the staff that had served him. A tough, hard drinking officer, Admiral Bowen was inclined to speak his mind.[7] He had already accumulated a number of high ranking enemies within the Navy. When confronted with scientists who assumed that they, and not he, knew what new weap-

[7] Interview.

ons the Navy needed, Admiral Bowen was not at all hesitant to acquire new adversaries outside the Navy. The sophisticated and determined men who rose to assert leadership in American science at the start of the Second World War—Vannevar Bush, then President of the Carnegie Institution and formerly Dean of Engineering and Vice President of the Massachusetts Institute of Technology (m.i.t.); James B. Conant, the President of Harvard University, and Karl T. Compton, then President of m.i.t.—had little patience for either the regular fellow, backslapping subculture of Washington or the formalities of military protocol. If they, as the administrators of elite academic research institutions, showed some disdain for those who made their careers in the military or a government laboratory, it was not totally unintended. Another officer acting as the Navy's liaison with the scientists and another group of representatives for the scientific community might have established a different set of relationships.

Admiral Bowen had not sought the position of Technical Aide to the Secretary. In 1939 when the assignment was made, he was hoping for a renewal of his appointment as Chief of the Bureau of Engineering.[8] During the preceding years he had forced the adoption of high pressure steam for the power systems of warships against the strong resistance of the Bureau of Construction and Repair. In 1939, the Naval Research Laboratory, which was then part of the Bureau of Engineering, began with his approval the exploration of nuclear fission as a potential power source for submarines.[9] At the same time, Admiral Bowen became embroiled in a bitter controversy with officers in the Bureau of Construction and Repair over the responsibility for a new class of destroyers that exceeded its design weight maximum. The Secretary of the Navy, Charles Edison, chose to resolve the dispute and avoid similar ones in the future by requiring a merger of the bureaus into a single entity to be known as the Bureau of Ships.[10] Bowen's several enemies within the Navy saw to it that he was removed from consideration for the chief of the new bureau.[11] His consolation prize was an assignment as Director of the Naval Research Laboratory (nrl), the jurisdiction for which was removed to the shelter of the Secretary's Office, a location that Secretary Edison's father

[8] Harold G. Bowen, *Ships, Machinery, and Mossbacks: The Autobiography of a Naval Engineer* (Princeton, N.J.: Princeton University Press, 1954), 120.

[9] Ibid., 183. See also Hewlett and Duncan, *Nuclear Navy*, 17–19, and Carl O. Holmquist and Russell S. Greenbaum, "The Development of Nuclear Propulsion in the Navy," *U.S. Naval Institute Proceedings*, September 1960, 65–71.

[10] Bowen, *Ships, Machinery, and Mossbacks*, 116–24. Memorandum, Secretary Charles Edison to file: "History of Reorganization Effort," 16 December 1939.

[11] Bowen, *Ships, Machinery, and Mossbacks*, 120–21.

(the great inventor Thomas Edison) had preferred when, as Chairman of the Naval Consulting Board during the First World War, he proposed the creation of the laboratory.[12] To save face for Bowen, the Secretary also appointed him as Technical Aide, a billet without many specific functions.[13] Once in this post, though, Admiral Bowen was anxious to use it to gain a role in the management of the Navy's research effort. He soon proposed the creation of a Naval Research Center built around NRL which would have bureau status and which would be responsible for the introduction of new technology.[14] As the civilian scientists were seeking the same role in both the Army and the Navy, their clash was inevitable.[15]

The strain in the relations between civilian scientists and the Navy, however, involved substantive issues as well as conflicting ambitions and personalities. Believing the United States should join in the war against Nazi Germany, Bush, Conant, and other prominent scientists wanted an early mobilization of the nation's technical manpower, but were apprehensive about the form such a mobilization would take if the initiative for its organization was left to the military. Fresh in their minds was the frustration of their own experience during the First World War. In that conflict, the military was initially reluctant to admit a need for outside assistance in the design of weapons, and then insisted on dominating the hurriedly created scientific effort that only began with the involvement of American troops in the fighting. Scientists who wished to contribute to the war by doing weapons-related research were required, with rare exception, to accept military commissions and to work at government facilities under military command procedures. Research priorities were determined by the military, and no attention was paid to linking weapon development to operational experience. Although some major advances in weapons were achieved, their impact on the outcome of the war was negligible beyond contributing to its frightful cost in lives.[16]

[12] The history of this venture in science advice and the laboratory is recorded in Lloyd N. Scott, *Naval Consulting Board of the United States* (Washington, D.C.: Government Printing Office, 1920); K. T. Compton, "Edison's Laboratory in Wartime," *Science* 75 (1932): 70–71; Carroll W. Pursell, Jr., "Science and Government Agencies," in D. D. Van Tassel and M. G. Hall, eds., *Science and Society in the United States* (Homewood, Ill.: Dorsey Press, 1966), 223–49. See also notes to chap. 6 below.

[13] Bowen, *Ships, Machinery, and Mossbacks*, 137.

[14] Memorandum to Secretary of the Navy from Director of Naval Research Laboratory, 29 January 1941.

[15] Interview.

[16] See I. Bernard Cohen, "American Physicists at War: From the First World War to 1942," *American Journal of Physics*, October 1945, 223–35; Kent C. Redmond, "World War II, a Watershed in the Role of the National Government in the Advancement of

Vannevar Bush was well situated to contest the terms of the scientists' involvement in the Second World War. In September of 1939, when the war broke out in Europe, he was already the Chairman of the National Advisory Committee for Aeronautics (NACA), a civilian agency which had been established in 1915 to promote the scientific development of aviation and which had been given authority (in June 1939) to assist the military in the development of combat aircraft. In June 1940, when allied resistance to the German blitzkrieg appeared to be disintegrating, Bush obtained presidential authorization to create and chair a parallel organization for promotion of military-related research in fields other than aviation, the National Defense Research Committee (NDRC).[17] In keeping with the parallel to the NACA, each of the services was given representation on the NDRC. Brig. Gen. George V. Strong represented the Army; Rear Admiral Bowen, the Navy.[18]

Using NDRC as his preferred forum, Bush quickly began to assert the scientists' independence from the military. He appointed a number of panels of leading academic specialists to reexamine the military's weapon development plans and to recommend opportunities for new avenues of research. He established a contract procedure for universities that permitted scientists to conduct weapons research in their own laboratories rather than in government laboratories and which compensated the universities for all costs incurred in the use of their facilities.[19] He also initiated contact with the British to create

Science and Technology," in C. Angoff, ed., *The Humanities in the Age of Science* (Rutherford, N.J.: Fairleigh Dickinson University Press, 1968), 166–80; Daniel J. Kevles, "Federal Legislation for Engineering Experiment Stations: The Episode of World War I," *Technology and Culture*, April 1971, 182–89; Robert A. Millikan, "Contributors of Physical Science," in John C. Burnham, ed., *Science in America* (New York: Holt, Rinehart, and Winston, 1971), 302–13; A. Hunter Dupree, *Science in the Federal Government* (Cambridge: Harvard University Press, 1957), 302–25; Gilbert F. Whittemore, Jr., "World War I, Poison Gas, Research and the Ideas of American Chemists," *Social Studies of Science*, May 1975, 135–63; Harvey M. Sapolsky, "Science, Technology and Military Policy," in I. Speigel-Rosing and D. de Solla Price, eds., *Science Policy Studies in Perspective* (London: Sage, 1977).

[17] Robert E. Sherwood, *Roosevelt and Hopkins: An Intimate History* (New York: Harper and Brothers, 1948), 154–56; Bush, *Pieces of the Action*, 30–37; Dupree, *Science in the Federal Government*, 369–71.

[18] Other members of the NDRC were Bush; Conant; Compton; Frank C. Jewett, President of the National Academy of Sciences; Conway P. Coe, Commissioner of Patents; and Richard C. Tolman of the California Institute of Technology.

[19] The importance of this step in changing government science relations is discussed in James B. Conant, *My Several Lives* (New York: Harper & Row, 1970), 234–47. See also Irwin Stewart, *Organizing Scientific Research for War: The Administrative History of the Office of Scientific Research and Development* (Boston: Little, Brown & Co., 1948); Bush, *Pieces of the Action*, 38–39; and notes to chapter 3 below.

an exchange of weapon designs and to gain information on their experience in countering German weapons.[20]

The military welcomed none of these steps. With American involvement in the war becoming more probable, the services were increasingly concerned about potential breaches in security. A free exchange of secrets with the British was thought too risky to undertake since Britain might soon be forced to surrender.[21] Similarly, the placement of weapon research contracts with universities was considered undesirable because the universities were centers of pacifist activities, and because university scientists were less likely than government employees to be security conscious. When it came to the NDRC panel examinations of the military's weapon development plans, the services saw a need not for the review of the plans, but rather for their fulfillment. Both services held these views, but because the Navy was the more jealous of its authority, being always the smaller, it was the service that was most likely to be the first to attempt to limit the autonomy of scientists.[22]

A confrontation between the Navy and civilian scientists could have occurred in a number of different fields. The Navy, for example, had been the first agency of the government to seek a military application for nuclear fission. In 1939 NRL, with Admiral Bowen's approval, had allocated funds to explore the potential of nuclear power for submarines.[23] Planning jurisdiction for nuclear matters was assigned to NDRC when it was created in 1940. With NDRC in charge, however, the government's attention in nuclear fission began to focus exclusively on the development of a bomb. The Navy, again through NRL, had pioneered the development of radar with its interest in the technology dating back to experiments conducted in the early 1920s. After reviewing progress in radar, however, NDRC chose in 1941 to emphasize the British microwave contributions and to concentrate its own work in that field at a new facility, the soon-to-be-famous Radiation Laboratory, to be located in Cambridge at M.I.T.[24]

[20] Conant, *My Several Lives*, 248–71.

[21] Ibid., 250 and Bush, *Pieces of the Action*, 21.

[22] The Army's suspicions of scientists is recorded in Constance M. Green, Harry C. Thomson, and Peter C. Roots, *The Ordnance Department: Planning Munitions for War* (Washington, D.C.: Government Printing Office, 1955), 226–32.

[23] Hewlett and Duncan, *Nuclear Navy*, 17–19.

[24] NDRC work on radar is discussed in Baxter, *Scientists Against Time*, 136–69. Cf. Bowen, *Ships, Machinery, and Mossbacks*, 138–81. The M.I.T. Radiation Laboratory is described in "Longhairs and Shortwaves," *Fortune*, November 1945, 163 ff. The NRL experience is described in David Kite Allison, *New Eye for the Navy: The Origins of Radar at the Naval Research Laboratory* (Washington, D.C.: Government Printing Office, 1981).

An earlier (and as it turned out, decisive) conflict arose over anti-submarine warfare, a field in which the Navy felt it had exclusive control. On 27 June 1940, the day that the NDRC was formally established, the Secretary of the Navy requested the National Academy of Sciences, an organization which was closely linked to the NDRC, to convene a committee to advise the Navy on scientific methods of detecting submarines and on the adequacy of the Navy's research program in antisubmarine warfare.[25] It was clear at the time that if the United States entered the war in Europe, the Navy would have to fight its way across an Atlantic Ocean dominated by U-boats. In establishing NDRC, the President had to ask the service secretaries for their support in utilizing the skills of civilian scientists to further defense preparations.[26] The request to the Academy made by the Secretary of the Navy was an indication of his willingness to cooperate with civilian scientists in this effort. It remained to be seen whether or not the uniformed Navy would also be willing to cooperate.

The Academy report, which stressed the need for a national commitment to intensify antisubmarine research, was submitted on 28 January 1941 to Admiral Bowen as the Secretary's Technical Aide.[27] In an elaborating letter sent to Bowen in early February, Frank Jewett, the President of the National Academy of Sciences (and like Bowen a member of NDRC, suggested that civilian scientists become involved in work on the submarine detection problem through the NDRC.[28] Over a month passed without a response from Bowen. Finally, after repeated inquiries from Jewett, the Admiral submitted the report to the Secretary with a memorandum rejecting its advice.[29] To Bowen, the scientists were "Johnny-Comelatelies."[30] The Navy already knew what was needed without being second guessed by outsiders. If civilian scientists wanted to be of assistance, they would have to work under the direction of plans established by naval officers.[31] "It would ap-

---

[25] Baxter, *Scientists Against Time*, 172.

[26] Ibid., 15.

[27] *Report of the Subcommittee on the Submarine Problem* (confidential), Navy Research Advisory Committee, National Academy of Sciences, Washington, DC, 28 January 1941.

[28] Frank B. Jewett letter to Adm. H. G. Bowen (confidential), 11 February 1941, National Academy of Sciences files.

[29] Rear Adm. Julius A. Furer, USN (Ret.), *Administration of the Navy Department in World War II* (Washington, D.C.: Department of the Navy, 1959), 776. See also William Bradford Huie, "The Backwardness of the Navy Brass," *The American Mercury* 62, no. 270 (June 1946): 647–53 which rehashes the Admiral Bowen story in an apparent attempt by the Army Air Force to influence postwar research plans.

[30] Bowen, *Ships, Machinery, and Mossbacks*, 178.

[31] Furer, *Administration of the Navy Department*, 776.

pear that at the present time that the only reason for acceding to such a recommendation [for independent participation by NDRC]," Bowen told the Secretary, "would be on account of the pressures exerted by certain well known scientists, some of whose names appear in this correspondence."[32]

Bush's name was not listed in the correspondence, but he obviously was one of the scientists Admiral Bowen had in mind and he soon came to the rescue of his colleagues. Bush knew that if scientists were to be treated as equals rather than as subordinates in their relations with the military, Bowen had to be disciplined. Bush's standing at the White House had risen rapidly since his appointment as chairman of NDRC. He had been impressive in the handling of nuclear matters and he had intervened to save the President from making what certainly would have been politically embarrassing appointments in medical research.[33] Already there were plans being formulated to give him executive authority through the establishment of an Office of Scientific Research and Development that would report directly to the president.[34] It was easy then for Bush to work his way around the obstinate Admiral Bowen. First, he persuaded the leadership of the Bureau of Ships, Bowen's arch rivals within the Navy, to invite NDRC's participation in antisubmarine warfare research.[35] Next, he had the Secretary of the Navy, now Frank Knox (Edison having been replaced in the previous year), appoint Professor Jerome C. Hunsaker of M.I.T., a former naval officer, to review the Navy's organization in research and the handling of the Academy report.[36]

The selection of Professor Hunsaker signaled Bowen's defeat, for Hunsaker was not only a naval constructor by training, and thus a member of the group within the Navy with whom Bowen had been in continuing conflict, but he was also Treasurer of the National Academy of Sciences, the organizational sponsor of the report whose value Bowen had challenged.[37] Hunsaker's recommendations, all of which were accepted by Knox, included the rejection of Admiral Bowen's plan for a Naval Research Center with bureau status, the transfer of NRL and Admiral Bowen to the Bureau of Ships, and the establish-

---

[32] Letter from the Director, Naval Research Laboratory to Secretary of the Navy, 17 March 1941; subj.: Report of Colpitts' Subcommittee on Antisubmarine Devices, quoted in Furer, *Administration of the Navy Department*, 776.

[33] Bush, *Pieces of the Action*, 43–44.

[34] Conant, *My Several Lives*, 272.

[35] Baxter, *Scientists Against Time*, 173.

[36] Furer, *Administration of the Navy Department*, 776.

[37] Hunsaker was also a pioneer in aeronautical engineering, having established the first department in the subject at M.I.T. while on loan from the Navy in 1913. L. Lord and A. D. Tumbrill, *History of Naval Aviation* (New Haven, Conn.: Yale, 1949), 23.

ment of an Office of the Coordinator for Research and Development for the Navy's liaison with civilian scientists.[38] Besides handing Bowen over to his enemies in the Bureau of Ships, Secretary Knox placed an unsatisfactory fitness report—Bowen's first—in his personnel file, the charge being that he showed poor judgment in his dealings with civilian scientists.[39] On 12 July 1941, the Secretary formally established the Office of the Coordinator of Research and Development and appointed Dr. Hunsaker as the coordinator to serve on a half-time basis.[40] Hunsaker would work with the Office of Scientific Research and Development established by the President on 28 June 1941, and headed by Vannevar Bush.[41]

The Academy report incident marked more than a personal defeat for Admiral Bowen. It was, as the official history of the administration of the Navy Department during the Second World War states, an incident that helped clarify the Navy's relationship with civilian scientists.[42] The disciplining of the Admiral was a clear indication of the support scientists had at the highest levels of government. Although military officers might continue to resent their participation as equals in the planning of weapon research, they could not openly resist such participation. For the duration of the war, at least, scientists had achieved the autonomy they had sought.

The Navy as well as Bowen was made to pay a price for not being appropriately deferential. Some months later, when Bush had to select between the services for the assignment of the project to build the atomic bomb, he preferred the Army rather than the Navy. Although many factors influenced the decision to place the Army in charge, Bush did not forget the past conflict with Admiral Bowen. As Bush subsequently explained, he did not think that "naval officers, especially those at the Naval Research Laboratory, had sufficient respect for and an ability to work cooperatively with civilian scientists" to be

---

[38] Hunsaker's report is discussed at length in Furer, *Administration of the Navy Department*, 776–80. Official letter from Dr. J. C. Hunsaker to Under Secretary of the Navy, 27 June 1941.

[39] Bowen, *Ships, Machinery, and Mossbacks*, 230. The report read: "Very capable engineer and possessed of great courage in pressing experimental projects. Rendering exceptional service in research work. Rather belligerent and temperamental in his contacts outside the Navy." Adm. Chester W. Nimitz, the Chief of Naval Personnel, tried to have the Secretary change the report, but failed.

[40] Department of the Navy, General Order 150, 12 July 1941 "Coordinator of Research and Development," reproduced in Furer, *Administration of the Navy Department*, 779–80.

[41] Baxter, *Scientists Against Time*, 124. See also Don K. Price, *Government and Science* (New York: New York University Press, 1954), 43–46.

[42] Furer, *Administration of the Navy Department*, 775.

given managerial responsibility for the most important weapon project of the war and one that would require close collaboration with scientists.[43]

THE BIRD DOGS PLAN THE POSTWAR RELATIONSHIP

Weapon research and development activities were not centrally managed in the United States during the Second World War. The four major agencies involved in these activities—OSRD, the War and Navy departments, and NACA—were organizationally independent of, and not always on friendly terms with, one another. OSRD, itself composed of a number of quasi-autonomous committees and divisions, made no attempt to implement its presidential charge to coordinate the research programs of the military services. Each of the Army's seven technical branches and the Navy's six materiel bureaus managed their own research, working as they chose, through affiliated government laboratories and arsenals or private contractors. NACA, which was closely linked to the Army Air Corps (the predecessor to the Air Force), had its own network of laboratories and contractors. Filling in the gaps left by these agencies as its leaders saw them, is what OSRD chose as its mission.

The Navy's Office of the Coordinator of Research and Development (the organization established on the recommendation of Professor Hunsaker, and initially headed by him) did not coordinate the research work of the Navy's materiel bureaus. Neither its small staff nor the bureaus would have permitted such a venture. Rather, it acted as a liaison agency between research organizations within the Navy and those outside, especially the civilian directed committees and divisions of OSRD.[44] In hundreds of projects, information had to be exchanged, equipment borrowed, and tests arranged. When new weapons were developed, the fleet had to be persuaded to accept them and learn how to use them.

Friction was inevitable in these relationships. Naval officers often

[43] Vincent Davis, *The Politics of Innovation: Patterns in Navy Cases* (Denver: Graduate School of International Affairs, University of Denver, 1967), 25. Davis cites correspondence from Bush to this effect in his book *The Admirals Lobby* (Chapel Hill, N.C.: University of North Carolina Press, 1967), 175 n. 45. See also his *Postwar Defense Policy and the U.S. Navy, 1943–1946* (Chapel Hill, N.C.: University of North Carolina Press, 1966), 338 n. 112. This opinion was confirmed in an interview that I had with Vannevar Bush, 23 February 1972.

[44] See Rear Adm. J. A. Furer, "Research in the Navy," *Journal of Applied Physics*, March 1944, 209 for a formal discussion of the duties of the office. Note also Organization Planning Staff, Navy Management Office, "History of the Office of Scientific Research and Development," 22 September 1961, processed.

felt that civilian scientists ignored their special knowledge and past achievements. Scientists, pleased with their civilian status, were usually unaccepting of the bureaucratic procedures of the Navy. Despite the urgency imposed by an approaching war, there were organizational interests on both sides to protect.[45] Hunsaker, as a naval officer converted to professor, had a special appreciation of the causes of the friction and the strong desire to minimize its impact on weapon research. He recruited into the Office of the Coordinator several young naval reserve officers with technical backgrounds to locate for him problems in interorganizational relations as they were developing. Hunsaker gave these officers a nickname, "Bird Dogs," which he felt described their function.[46]

Shortly before the attack on Pearl Harbor, Hunsaker became chairman of NACA, selecting Rear Adm. Julius Furer, a naval constructor who was nearing retirement, as his successor in the now full-time post as Coordinator of Research and Development in the Navy.[47] Because Admiral Furer was a less assertive leader than was Professor Hunsaker, the Bird Dogs gained more independence within the Office of the Coordinator and gradually began to acquire side interests in addition to their regular duties.[48] The side interest that attracted several of them, especially Lts. Bruce S. Old and Ralph A. Krause, was the future organization of research activities within the Navy—a large undertaking for young officers, reservists at that.[49]

In December 1942, Old and Krause began a series of monthly evening meetings to discuss the research the Navy would need after the war and how it might be structured. They were concerned that the civilian interest in naval relevant science and technology as demonstrated in the work of OSRD would fade in peacetime though the Navy's need for superiority in new equipment would not. Industry, tied

[45] Interviews.

[46] The Bird Dogs, "The Evolution of the Office of Naval Research," 31.

[47] General Order No. 150 of 12 July 1941 which had established the Office of the Coordinator of Research and Development was modified by General Order No. 159 of December 1941 to permit naval officers as well as civilians to hold the post of coordinator. One reason suggested for the selection of Admiral Furer was that Hunsaker knew that he had no real career left in the Navy and thus thought that he would be an aggressive leader. Events proved otherwise. Interview.

[48] Admiral Furer may have been preoccupied with an additional assignment of monitoring nuclear affairs for the Navy. Interview.

[49] Although junior officers during the Second World War, several of the Bird Dogs became important figures in science and the management of research during the postwar years. Old, for example, became a Vice President of the A. D. Little Co., in Cambridge, Massachusetts and Krause was President of Stanford Research Institute, formerly part of Stanford University and now an independent organization.

in peacetime to the pursuit of profits, could not be expected to fill the void that OSRD would leave. The Navy, they thought, had to provide the organizational framework to stimulate continual progress in science and technology relevant to its needs.[50] Joining them at their first session as a senior adviser was George B. Karelitz, a professor of mechanical engineering at Columbia University and a former officer in the Imperial Russian Navy, who had had a distinguished career in industrial and academic research.[51] Although Karelitz died unexpectedly the next month of a heart attack, the Bird Dogs persisted in their quest.[52] By November 1943, they had developed an organizational design that they felt was adequate for the task.[53]

The design called for the appointment of a line officer with the rank of Rear Admiral as director of [naval] research to report to the Secretary of the Navy and to be responsible for the support and management of university-based research of relevance to the service. The choice of a line officer with flag rank was to assure good working relations with the fleet; the location of his organization in the Office of the Secretary was to give him some protection from the fleet and equal status with the chiefs of the materiel bureaus. Assisting the Director and actually managing the organization's research program were to be two civilian deputies, each of whom was to have extensive academic or industrial research experience. Two committees were to provide policy guidance. One, to be known as the Research Advisory Board, was to be composed of leading scientists and representatives of the Army and NACA; the other, the Naval Research and Development Board, was then already in existence and was composed of officers representing research organizations and staffs within the Navy.[54] A later version, prepared after further discussions during the next year, included a proposal for the appointment of an Assistant Secretary of the Navy for Research to coordinate the department's research activities.[55] With but minor modifications, this was the organizational design for research the Navy was eventually to adopt.[56]

---

[50] Minutes, first meeting, Postwar Research Planning Group, 14 December 1942. B. S. Old files.

[51] Interview.

[52] Obituary, *New York Times*, 20 January 1943.

[53] The Bird Dogs, "The Evolution of the Office of Naval Research," 32.

[54] Organizational Chart, RAK-BSO, 1 November 1943, Interview.

[55] Draft Memorandum from Lt. Comdr. R. A. Krause, U.S.N.R.; Lt. Comdr. Bruce S. Old, U.S.N.R.; and Lt. John T. Burwell, Jr., U.S.N.R. to Secretary of the Navy via the Coordinator of Research and Development. Subj.: Suggestion for Post-War Organization of Naval Research. 23 September 1944.

[56] The main modifications both incorporated in later draft designs prepared by the Bird Dogs were the elimination of the requirement that the Director of Naval Research

Old and Krause, then Lieutenant Commanders, joined by Lt. John T. Burwell (another Bird Dog in the Office of the Coordinator of Research and Development), decided in September 1944 to submit their plan directly to the Secretary of the Navy as a "Beneficial Suggestion."[57] Naval Regulations permit anyone in the Navy to prepare such a suggestion and require commanding officers, whether or not they approve, to forward them. When Admiral Furer received the Bird Dogs' Beneficial Suggestion for transmittal to the Secretary, he was quite upset, for he thought their use of this device indicated a lack of faith in his leadership by officers under his command.[58] The Bird Dogs had not wanted to involve the Admiral in their plans because they knew that he was not liked by the Secretary and feared that he would claim their ideas for the organization of postwar research as his own. Reluctantly, they agreed to have him officially sponsor the suggestion.[59] The Secretary, however, chose to ignore it.[60]

The failure of the plan to arouse the interest of the Secretary freed the Bird Dogs from the obligation of working through Admiral Furer. From junior reserve officers on the staff of the Chief of Naval Operations, they learned that President Roosevelt wanted to retain personal control over the structure of the Navy. This was manifested by the fact that the President had recently returned a proposal for a minor reorganization from Adm. Ernest J. King with a handwritten marginal note stating, "Ernie, you worry about fighting the war; I'll worry about the organization of the Navy, FDR."[61] Using the same network of contacts, the Bird Dogs managed to have themselves scheduled in April 1945 for an appointment with the President on his expected return to Washington from Warm Springs, in order to present their proposal personally to him.[62] Roosevelt, however, died before the appointment took place. Within a month, Secretary Forrestal ordered that the Office of the Coordinator be absorbed by another agency, the Office of Patents and Inventions. The combined agency, titled the Office of Re-

---

be a line officer and that there be co-equal civilian assistants. The title "Director" was changed to "Chief" in the legislation establishing ONR so as to be less offensive to the Chiefs of the Bureaus, "Director" implying more formal authority over naval research than "Chief." Interviews and Organizational Chart for the Office of the Assistant Secretary of the Navy for Research, 23 September 1944, initialled RAK, BSO & JTB.

[57] Interview.

[58] Interview.

[59] Memorandum from the Coordinator of Research and Development to the Secretary of the Navy. Subj.: Organization of Research in the Navy Department. 11 October 1944.

[60] Interviews.

[61] Interview.

[62] The Bird Dogs, "The Evolution of the Office of Naval Research," 35.

search and Inventions, was to be headed by Rear Admiral Bowen. The Bird Dogs' effort to plan the organization of the Navy's postwar research program seemed at an end.[63]

THE RETURN OF ADMIRAL BOWEN

After his defeat by Vannevar Bush, Admiral Bowen had to endure a year as a subordinate official in the Bureau of Ships until his tour as Director of NRL was over.[64] His humiliation was not yet complete, for his enemies within the Navy barred him from a major command and nearly caused his retirement by a medical review panel. Only a growing friendship with the Under Secretary of the Navy saved him from being forced to collect an early pension in the midst of the greatest war in the Navy's history.[65] First on a part-time basis and then full-time, Admiral Bowen had been assisting the Under Secretary in arranging seizure for the government of munitions factories and shipyards, where production was threatened by labor disputes. His toughness in these difficult situations won him the gratitude of the Under Secretary, James Forrestal.[66] With the death of Secretary Knox in the spring of 1944, Forrestal became the Secretary of the Navy. Soon Forrestal had another opportunity to express his gratitude for Admiral Bowen's aid.

Forrestal was Secretary in November 1944 when the Coordinator of Research and Development, Admiral Furer, submitted in his own name the Bird Dogs' design for the organization of the department's research effort. On Forrestal's desk at the time, as it had been since he became secretary, was a report on patent policy prepared by R. J. Dearborn, an industrialist working for the Navy during the war, and also a friend of Admiral Bowen.[67] The report contained a proposal for the establishment of an office to oversee the Navy's patent applications. Infringement claims amounting to millions of dollars had been filed against the Navy after the First World War, and though only a few (including some pressed by the Royal Navy) were successful, the

---

[63] Ibid., 35 and interviews.

[64] Bowen, *Ships, Machinery, and Mossbacks*, 136.

[65] Ibid., 231.

[66] Ibid., 205–344; see also R. G. Albron and R. H. Connery, *Forrestal and the Navy* (New York: Columbia University Press, 1962). "Two-Fisted Admiral Bowen, Boss of Navy-Seized Shops, 'Impartial as Meat-Ax,'" *San Francisco Chronicle*, 5 September 1944, Bowen Papers, Clipping File, Box 11, Princeton University.

[67] R. J. Dearborn, "Proposed Policies and Procedures of the Navy Department in Respect of Patents and Inventions," 10 March 1944 (Report to the Secretary of the Navy); Bowen, *Ships, Machinery, and Mossbacks*, 346–47. Admiral Bowen nominated Dearborn for the Navy Distinguished Service Award at the end of the war.

legal defense against them had been costly. Dearborn's recommendations were intended to avoid a recurrence of this situation after the end of the Second World War.[68] Eight days after receiving the Bird Dogs' design, Secretary Forrestal decided to implement Dearborn's now somewhat dusty proposal by establishing an Office of Patents and Inventions and appointing Admiral Bowen as its Director.[69]

The title bestowed upon Admiral Bowen belied the Secretary's real intention. Instrumental in his decision to create the Office of Patents and Inventions and to select Bowen as its director, was Commodore Lewis L. Strauss, an investment banker with a reserve commission and later a member and controversial chairman of the Atomic Energy Commission.[70] At the time, Strauss was Forrestal's Special Assistant and his key aide for research planning and nuclear affairs.[71] Like Forrestal, Strauss admired Admiral Bowen's boldness in promoting innovation and skill in managing research.[72] To those in the Office of the Coordinator of Research and Development, the establishment of the Office of Patents and Inventions, regardless of its formal duties, was intended to give Admiral Bowen a role in the planning of the Navy's postwar research strategy, a task which they had hoped would be exclusively their own.[73]

Forrestal did nothing to clarify the conflicting organizational jurisdictions until the spring of 1945 and the death of President Roosevelt. On 19 May 1945, he ordered the creation of an Office of Research and Inventions (ORI) and the transfer to it of all the functions of both the Office of Patents and Inventions and the Office of the Coordinator of Research and Development as well as jurisdiction over the Naval Research Laboratory and postwar research planning; Admiral Bowen

---

[68] R. J. Dearborn, "Proposed Policies," 3.

[69] Admiral Furer, shortly before submitting the Bird Dogs' proposal, attempted to argue that Dearborn's ideas could best be implemented in a central research organization. Department of the Navy, Office of the General Counsel, Patent Division, Memorandum to the Secretary of the Navy reporting meeting held on 25 September 1944, dated 28 September 1944; the Office of Patents and Inventions was established by a directive signed by the Secretary of the Navy on 19 October 1944.

[70] Interview. Nuel Pharr Davis in a book not very favorable toward Strauss does include the following judgment: "Genuinely humble, genuinely in love with science—the Office of Naval Research, the best of the government's science-subsidy programs was his child—Strauss did not think of nuclear weapons as a curse," *Lawrence and Oppenheimer* (New York: Simon and Schuster, 1968), 278. Strauss discussed the ONR in his autobiography, *Men and Decisions* (New York: Doubleday, 1962), 146–48. See also Richard Pfau, *No Sacrifice Too Great: The Life of Lewis L. Strauss* (Charlottesville: University of Virginia Press, 1984), 77.

[71] "Lewis L. Strauss, Ex-A.E.C. Head, Dies," *New York Times*, 22 January 1974.

[72] Interview.

[73] Interviews.

was appointed its Chief.[74] Four years after his defeat, Admiral Bowen was once again dominant. He was in charge of an office that absorbed the office established to placate the scientists he had offended. With Admiral Bowen officially in command of the Navy's postwar research planning, the Bird Dogs thought that their vision of a Navy supporting the work of civilian scientists had been totally shattered. They were to be surprised.[75]

Admiral Bowen apparently intended to use ORI as the agency to promote the development of nuclear propulsion for the Navy.[76] His first objective in that quest was to gain access for the Navy to the Manhattan District and current progress in nuclear physics. Barring access was the Army, content with its atomic bomb monopoly. Requests for atomic clearances for naval officers in the early months after the end of the war were put off by Gen. Leslie Groves, the Army officer heading the Manhattan Project.[77] Protests, even those submitted through Secretary Forrestal, failed to budge General Groves.[78] However, not all doors to nuclear information were well guarded. Admiral Bowen soon discovered that General Groves had alienated many of the civilian scientists who had been mobilized for the bomb project with the strict discipline that he had imposed upon their work. Dissatisfied with Army management, they were streaming back to the universities at the end of the war, although still anxious to continue their research.[79]

[74] Letter from the Secretary of the Navy to all Bureaus, Boards, and Offices of the Navy Department. Subj.: Office of Research and Inventions, 19 May 1945. Also, "Navy Sets Up Office of Research and Inventions," Chemical and Engineering News 23 (10 July 1945): 1159.

[75] The Bird Dogs, "The Evolution of the Office of Naval Research." Interviews.

[76] Interviews. See also Hewlett and Duncan, Nuclear Navy, chapter 2.

[77] Carl O. Holmquist and Russell S. Greenbaum, "The Development of Nuclear Propulsion in the Navy," U.S. Naval Institute Proceedings, September 1960, 68. Bowen in a letter to Capt. Martin Lawrence (27 July 1951) said that the only omission of significance in his autobiography was the struggle with General Groves for access to atomic energy Bowen Papers, Box 1, Princeton University.

[78] Bowen first requested clearances for ORI in late December 1945. When a negative reply was finally received from General Groves on 26 February 1946, Secretary Forrestal was solicited to formally protest the failure to give the Navy access. Letter from Secretary of the Navy Forrestal to Secretary of the Army Patterson, dated 16 March 1946, cited in R. S. Greenbaum, "Nuclear Power for the Navy: The First Decade," Office of Naval Research, March 1955, C Title U, Office of Naval History files. Reply letter from Secretary Patterson to Secretary Forrestal, dated 2 April 1946 and counter reply letter from Secretary Forrestal to Secretary Patterson, dated 20 April 1946. Naval History Center files.

[79] "Secrecy Hampers Atomic Research," New York Times, 3 February 1946; "Atomic Control by Army Decried," New York Times, 18 March 1946. Bowen was quick to seize upon this public criticism of the Army by arguing in favor of "civilian" control of nuclear

Formerly abrupt with civilian scientists, Admiral Bowen became solicitous of their welfare, seeing them now as potentially useful allies rather than interfering outsiders as in the past.

Scientists were among the publicly recognized heroes of the war; their opinions, as Bowen was aware, were being sought and offered on many postwar issues, especially those relating to atomic energy and military affairs. To the amazement of the Bird Dogs, Admiral Bowen had immediately set about implementing their plan to make the Navy a patron of academic science. Under his orders, Capt. Robert Dexter Conrad, head of ORI's Planning Division, was sent across the country to persuade reluctant university presidents to accept Navy research contracts for their scientists.[80] Although the focus of the program was nuclear physics, money was available for the support of basic research in nearly every field of science.[81] At the same time, Admiral Bowen (through the NRL, which was once again under his direction) was preparing to move ahead with the project to develop nuclear propulsion for ships, whether or not an agreement could be worked out with General Groves.[82] It was likely that Admiral Bowen envisioned the day when the Navy would need the support of the scientists it was so graciously supporting at his behest.

The turnabout was incredible. Only three years before, Admiral Bowen had described NDRC's effort to provide research funds to universities as a "conspiracy" to subsidize academics.[83] Now he was conspiring to do the same.

---

research. "Admiral Pictures an 'Atomic' Navy," New York Times, 30 March 1946. Interview. See also n. 29 below.

[80] Memorandum from Capt. R. D. Conrad and Lt. Comdr. B. S. Old to Chief of Office of Research and Inventions. Subj.: Report of Visit to the University of Chicago, University of California, and the California Institute of Technology (n.d.). This series of visits occurred between 10 October and 21 October 1945. Conrad, once remembered for his opposition to basic research, became its advocate as an aide to Admiral Bowen. Interview. Capt. Conrad died prematurely of cancer and the Navy later named a research award and an oceanographic research vessel after him, the USNS Robert D. Conrad (AGOR-3) which is operated by the Lamont-Doherty Geological Observatory of Columbia University. For the opposition to military sponsorship, note "Yale Head Assails Controlled Study," New York Times, 4 January 1946, 23.

[81] The program is reported in John E. Pfeiffer, "The Office of Naval Research," Scientific American, February 1949, 11–15 and described officially in Office of Research and Inventions, Fundamental Scientific Research Program (1946?), NAVEXOS-P-354. Bowen noted in a report to the Secretary that although this work is basic, it will be useful for the development of nuclear propulsion by the Navy. Memorandum from Chief Office of Research and Inventions to Secretary of the Navy, 9 April 1946.

[82] Greenbaum, "Nuclear Power for the Navy," 22, and Hewlett and Duncan, Nuclear Navy, 17–19.

[83] Letter from Admiral Bowen to Sen. Peter G. Gerry, 23 May 1942. Bowen Papers, Box 1, Princeton University.

Aside from the Army, two obstacles remained in the Admiral's path. One was to obtain congressional legitimacy for his organization. Because ORI had been established by Executive Order under the War Powers Act, there was some question as to whether it could continue to commit public funds after the war had ended.[84] Moreover, several bills describing the Navy's postwar research structure had already been submitted.[85] If Bowen had not seized the initiative, it was quite possible that someone else would have. Finally, he knew as he expressed it, "In Washington, one cannot accomplish anything without a statute in front of him and an appropriation behind him."[86] Old and Krause, now working for Admiral Bowen, prepared the bill for what was to be called the Office of Naval Research (the Office of Research and Invention as a title was thought to evoke too much of a Rube Goldberg image)[87] the design they selected was the one they themselves had outlined two years before.[88] Helpful in the negotiations for the bill's passage were Lewis Strauss, always a power behind the scenes; W. John Kenney, a skillful Washington lawyer and then the Under Secretary of the Navy; and Admiral Luis de Florez, a flamboyant aviator and inventor, ORI's Deputy Chief, and later director of research at the Central Intelligence Agency.[89] Before the summer recess of the 79th Congress in August 1946, Admiral Bowen had the authorizing legislation he had sought.[90]

The remaining obstacle, the one Admiral Bowen never did overcome, was the need to gain the internal Navy jurisdiction, or cognizance as is preferred in the Navy, for the development of nuclear propulsion. Bowen's aggressiveness once again aroused opposition within the Navy. Secretary Forrestal, preoccupied with the developing battle

[84] Chairman Vinson of the House Naval Affairs Committee who had warned the Navy that the Executive Order establishing ORI was insufficient authorization in peacetime made the Navy admit its error in the hearings for ONR. *Hearings on H.R. 5911*, Committee on Naval Affairs, House, 1st Session, 79th Congress, 26 March 1946, 2835.

[85] Office of Naval Research, *The History of United States Naval Research and Development in World War II* 6 (n.d.), chap. 24.

[86] Bowen, *Ships, Machinery, and Mossbacks*, 351.

[87] Interview.

[88] The Bird Dogs, "Evolution of the Office of Naval Research."

[89] Admiral de Florez was an amazing individual thoroughly deserving of the *New Yorker* profile which appeared in the 11 and 18 November 1944 issues of the magazine. He is credited, for example, with inventing the first accurate antiaircraft sight in 1914. He worked as manager of an English factory at the time, and grew tired of German aircraft disrupting his work. Around M.I.T. he is remembered for arriving via seaplane on the Charles River on his visits to his alma mater.

[90] Public Law 588, 79th Congress. Letter from Secretary of the Navy. To: All Bureaus, Boards, and Officers of the Navy Department. Subj.: Office of Naval Research, 2 August 1946.

over unification of the armed services, apparently could offer little assistance this time.[91]

On 13 November 1945, the Chief of Naval Operations set up a planning office for nuclear weapons and guided missile developments—the Deputy Chief of Naval Operations for Special Weapons (OP-06)—despite Bowen's protests that the new office conflicted with his own duties as Chief of ORI. Bowen argued that these duties required that he alone coordinate naval research activities.[92] The Bureau of Ships, Admiral Bowen's old nemesis, had by then already indicated that it wanted the responsibility for the development of nuclear propulsion. Not surprisingly, the bureau's leadership could find no reason to oppose the establishment of OP-06, but many reasons to oppose Bowen's assertion of a coordinating role in naval research.[93]

The officers who staffed OP-06 were primarily those few naval officers that had had some contact with the Manhattan District during the war. It was their assessment that the Navy could never expect to be able to conduct an independent development of nuclear propulsion, as Admiral Bowen was threatening to attempt to do through NRL and ONR, because national policy would never permit decentralized control over nuclear research, a judgment with which the Bureau of Ships readily concurred.[94] Soon an alliance between OP-06 and Bureau of Ships officers isolated Bowen in the internal Navy discussions of nuclear matters and in the delicate negotiations with the Manhattan District over clearances for naval personnel. By the time President Harry S Truman signed the legislation establishing ONR on 3 August 1946, the jurisdictional issue within the Navy was already settled; Bowen had once again lost. It would be an officer the Bureau of Ships would designate (Captain Hyman G. Rickover it turned out), and not Admiral Bowen, who would direct a collaborative project with the Atomic Energy Commission (the successor agency to the Manhattan District) to develop nuclear propulsion systems for naval vessels.[95] The Office

[91] Forrestal's activities at the time are discussed in Demetrius Caraley, *The Politics of Military Unification* (New York: Columbia University Press, 1965) and Albion and Connery, *Forrestal and the Navy*, 242, 250–86.

[92] Letter from the Chief of Naval Operations to All Bureaus of Offices, dated 13 November 1945. OP-06-og Ser 8pog. A memorandum to the Assistant Secretary of the Navy in response to the CNO letter was prepared by Bowen's office in late November. Draft dated 24 November 1945. Subj.: Special Weapons Division of Naval Operations, B. S. Old files.

[93] Interviews; Hewett and Duncan, *Nuclear Navy*, 26; memorandum from Chief of the Bureau of Ships to Deputy Chief of Naval Operations (OP-06) 29 March 1946, Office of Naval History files.

[94] Hewlett and Duncan, *Nuclear Navy*, 26–27.

[95] Ibid., 28–38; Greenbaum, "Nuclear Power," 73–74; Interview.

of Naval Research, Bowen's secure organizational base for the creation of the nuclear navy, was essentially an organization in search of a mission from the day it was created. Admiral Bowen went on terminal leave shortly after ONR was established, although his official retirement date was to be listed as 1 June 1947.

THE SCIENTISTS FAIL TO GAIN POSTWAR INDEPENDENCE

Well before the end of the Second World War, the postwar organization of science was a topic on the agendas of the agencies that were managing the war's research effort. One spur to their interest in this topic was the drafting of legislation in 1943 under the sponsorship of Senator Harley M. Kilgore, Democrat of West Virginia, that called for the centralization into a single agency of all or at least most governmental research at the war's end. Although neither the military nor the wartime leadership of science thought that the Kilgore scheme was feasible or even likely of adoption, they were fearful that the populist-oriented senator and his socialist-inclined staff (some believed it to be communist dominated) would seize the initiative in planning for postwar science unless more practical and desirable alternatives were generated.[96] Another spur was the absurdity of the military's own initial venture into postwar planning for science. In late 1943 the Army's Ordnance Corps and the Navy's Bureau of Ordnance jointly put forward a proposal, which Lewis Strauss later claimed to have originated, that called for the establishment of a billion dollar fund, the capital of which would be invested in government bonds and whose annual income would be used to support ordnance-related research.[97] The attraction of this idea to its sponsors was a continuing source of support for ordnance research that would not be subject to the predictable postwar miserliness of Congress. Needless to say, the billion dollar fund proposal did not survive long once it was exposed to the criticism of more politically sophisticated officers and of Bush and his colleagues.[98]

More important, perhaps, were Bush's often repeated announcements, beginning in early 1944, that OSRD would be disbanded as soon

[96] Interviews. Memorandum from Lebrand Smith to R. A. Furer. Subj.: Comments on Kilgore Bill S.702, 19 March 1943.

[97] Interview. See also Lewis L. Strauss, *Men and Decisions*, 146–48.

[98] Interview. Memorandum from R. A. Furer to R. A. Blandy. Subj.: Postwar Research, Comments on Army/Navy Ordnance Research Plan, 13 December 1943. B. S. Old files. Memorandum for Special Plans Division to Chief of Staff [U.S. Army], 14 February 1944. B. S. Old files.

as war conditions permitted.[99] Bush had always believed that scientists would not tolerate mobilization a moment longer than they could be convinced of its necessity. By the third year of the war and his fifth year in Washington, Bush himself was tiring of the jurisdictional squabbles spawned by ad hoc arrangements. He was particularly annoyed by the Navy, as it continued to distrust OSRD and civilian scientists. In the fall of 1944, when the war in Europe seemed headed for a quick resolution, Bush was ready "to phase down" OSRD. There was no point in trying to involve OSRD in the war in the Pacific, he wrote Admiral Furer, since the war in the Pacific was the Navy's war and "the Navy does not yet . . . fully grasp how to manage and collaborate with civilian scientific personnel either in the theaters or at headquarters."[100] It was time, he felt, for both the scientists and the military to focus their thinking on their long run relationships.[101]

The formal process of considering postwar plans for military research had actually begun months earlier at a conference of senior military officers and scientists who were involved in the management of the wartime effort. Although the conference had been convened nominally at the request of the Army, it was in fact Bush's show. He deftly ignored criticism by some of the officers present of scientists who refused direction. Instead, he concentrated attention on the need to keep scientists involved in weapons-related research and on the past difficulty in peacetime of obtaining sufficient resources for their work.[102] The stage-managed result of the conference was the establishment by the service secretaries of a committee to examine possible schemes for continuing the involvement of civilian scientists in weapons-related research in the postwar period.[103]

Chaired by Charles E. Wilson, then Executive Vice Chairman of the War Production Board, the Committee on Postwar Research was composed of four representatives each from the Army, the Navy, and the civilian scientific community.[104] Though the committee explored briefly a number of schemes during the summer of 1944, it concerned

[99] Stewart, *Organizing Scientific Research for War*, 299–300.

[100] Letter V. Bush to R.A.J. Furer, 30 September 1944. B. S. Old files.

[101] Letters notifying OSRD components and division of Bush's intention to disband the organization were distributed about this time. B. S. Old files.

[102] Transcript, meeting on Postwar Research, 26 April 1944. B. S. Old files.

[103] The committee was established on 22 June 1944. J. Furer, *Administrative History*, 801.

[104] In addition to Wilson, the committee members were civilians: F. Jewett, President of the NAS; J. Hunsaker, Chairman of NACA; K. Compton, OSRD; M. Tuve, OSRD; Army: General Echols, Army Air Corps; General Tompkins; General Waldron; General Osborne; Navy: Admiral Furer; Admiral Hussey, Bureau of Ordnance; Admiral Cochrane, Bureau of Ships; Admiral Ransey, Bureau of Aeronautics.

itself primarily with two that involved the creation of a Research Board for National Security (RBNS). One, offered by Frank Jewett, the President of the National Academy of Sciences, called for the placement of the RBNS within the Academy with its financing to be obtained through the earmarking of military appropriations. The other, presented by Merle Tuve (a pioneer in the development of the proximity fuse and a critic of the Academy), called for an independent RBNS with financing to be obtained through direct appropriations from Congress. As both plans were based on the same organizational design—a forty-member board to be divided equally between officers and civilians and a civilian controlled executive committee—their discussion centered on the relative merits of an Academy affiliation.[105]

Tuve argued that the earmarking of funds as required in the Academy plan would lead to military control of research, whereas his plan for an independent RBNS would not. He noted also that another branch of the Academy, the National Research Council, had been created during the First World War to perform functions quite similar to those envisioned for the RBNS under similar arrangements, but that it had grown moribund during the intervening years.[106] Jewett countered by claiming that an Academy affiliation would protect the RBNS from both the vicissitudes inherent in the legislative process to gain its enactment, and the likely later attempts to subjugate it to political control. Unnoticed by the partisans of both plans was a comment by Jerome Hunsaker, Chairman of the NACA, that in essence implied that their debate was irrelevant. In his own organization (supposedly civilian controlled), he told the committee that a nod by General Henry H. Arnold, the commanding officer of the Army Air Forces, always determined the fate of a research project without a vote ever taking place. Academy affiliation or not, the unstated message was that the military would dominate RBNS as long as the military participated in its activities.[107]

The action of the Wilson committee would itself soon support Hun-

105 Minutes, meetings of the Committee on Postwar Research, B. S. Old files. For a useful analysis, see Daniel J. Kevles, "National Science Foundation and the Debate over Post-war Research Policy," ISIS 68 (1977): 5–26.

106 On the origins of the National Research Council, see Hunter Dupree, Science and the Federal Government (Cambridge: Harvard University Press, 1960).

107 Notes, subcommittee meeting of the Committee on Postwar Research, 8 September 1944, B. S. Old files. One other comment by an anonymous member was equally interesting. He said that the President of the Academy would make an ideal chairman of the RBNS "except that a pacifist or a biologist might one day hold this office." The NACA's experience is thoroughly analyzed in Alex Roland's fine history, Model Research: The National Advisory Committee for Aeronautics, 2 vols. (Washington, D.C.: National Aeronautics and Space Administration, 1985).

saker's warning. For the military representatives on the committee, the only real issue was the likely impact of the plans on research budgets within the services. The problem was that research was differently situated in each of the services. The technical branches were in a relatively weak position within the Army, having had difficulty in the prewar years in gaining significant allocations. Their representatives on the Wilson committee preferred an Academy affiliation for the RBNS because they believed that the earmarking of funds would increase their budget shares within the Army without altering their intended programs. The Navy's materiel bureaus, on the other hand, saw earmarking as a potential threat as they had long been a powerful force within the Navy. Their representatives, not surprisingly, favored a formally independent RBNS with its own appropriations.[108] Despite the fact that a majority existed on the committee for an independent RBNS, the committee in the end endorsed both plans: the Academy RBNS as an interim solution and an independent RBNS as a long-term goal. With the armed services divided, a choice between the plans could not be made.[109]

The Wilson committee report, such that it was, was submitted in September 1944. Months passed as the services sought to gain adoption of their own particular interpretation of its recommendations.[110] Finally, in February 1945, with legislation for an independent RBNS soon to be introduced, Secretary of War Henry L. Stimson and Secretary of the Navy Forrestal established the "interim" RBNS at the Academy.[111] Complications developed immediately. Bush, for example, hinted publicly that the RBNS might not be enough; scientists needed independence from the military, he argued, if they were to push ahead

[108] Sanford L. Weiner, "Navy and the Scientists: A Case Study of Two Post War Experiences," MIT/ONR History Project, 15 January 1970. B. S. Old notes Committee on Postwar Research.

[109] The report and its recommendations are summarized in the testimony of Major General Tompkins to the Woodrum Committee; see next footnote. Also, "Postwar Military Research," *Science*, 24 (November 1944): 461.

[110] The Navy believed that the Army had broken the Wilson agreement in its testimony on the report when it called for a long-term test of the interim arrangement. Note statement presented by Maj. Gen. W. F. Tompkins before the Select Committee on Military Policy, 21 November 1944. Also memorandum, 30 November 1944, from Lt. H. G. Dyke to Coordinator of R&D, Subj.: RBNS, Army Conference on Procedures, held 27 November 1944.

[111] Senator Byrd of Virginia introduced the bill (S. 825) at the request of the Navy, on 4 April 1945. Drafting began the previous month. Memorandum from Lt. C.V.S. Roosevelt, U.S.N.R., to Commodore Strauss. Subj.: Your Comments Made on 30 July at Meeting No. 1 of the Navy Department Research Groups; "Research Board for National Security," *Science* 101 (March 1945): 226–28. Karl Compton was appointed its first and only chairman.

rapidly with new ideas.[112] In Congress, the bill to establish the board was fast becoming intertwined with the literally dozens of other bills relating to postwar research that were then being presented. Moreover, several committees in Congress were objecting to the existence of an "interim" RBNS pointing out that any transfer of military appropriations to a "private" organization like the Academy without explicit congressional authorization was illegal.[113] Worse yet, Harold D. Smith, the Director of the Bureau of the Budget, was claiming that an important governmental responsibility as the planning of military research could not be vested in a nongovernmental board without diminishing the constitutional authority of the President.[114] Smith soon backed this claim with a directive from President Roosevelt instructing the Secretaries of War and Navy to retain within the government full control over public resources and responsibilities. At the most, the secretaries were told, the RBNS could be advisory to the armed services in the performance of their official duties.[115] By fall of 1945, the services were quietly backing away from the RBNS in either form. By the beginning of the new year, the concept was dead.[116]

Bush had already opened a second front. In November 1944, he arranged to be requested by President Roosevelt to examine the ways in which science, so effective in aiding the war effort, might contribute to the peace.[117] The initiating stimulus was the likely introduction in the next session of Congress of yet another version of Senator Kilgore's bill, this time one that would focus on government-university relations.[118] The product of the Bush directed study was the famous report, *Science: The Endless Frontier*, which was submitted to President Truman in July 1945, and the legislation that would lead eventually (along with Kilgore's), to the establishment of the National Science Foundation (NSF). Never far from his thoughts, however, was the problem he perceived that scientists would have in the postwar period in gaining independence from the military in military-related research. Included in the legislative proposal for the civilian-directed

---

[112] *New York Times*, 25 February 1945, 8.

[113] Kevles, "National Science Foundation."

[114] Letter from the Director of the Bureau of the Budget to the Secretary of the Navy, 7 April 1945. ONR files. Kevles, "National Science Foundation," 25.

[115] Daniel J. Kevles, "Scientists, the Military, and the Control of Postwar Defense Research: The Case of the Research Board for National Security, 1944–46," *Technology and Culture* 16 (1975), 20–47.

[116] Letter from Secretaries Patterson and Forrestal to F. Jewett, President, National Academy of Sciences, 18 October 1945. B. S. Old files.

[117] The full story is outlined in a letter to the Editor of *Science* by Daniel J. Kevles, "FDR's Science Policy," *Science* 183 (1 March 1974), 798, 800.

[118] Milton Lomask, "The Birth of NSF," *MOSAIC*, November/December 1975, 23.

National Research Foundation (the title Bush proposed for the agency that became the NSF) was a provision establishing within the Foundation a Division of National Defense to support military research conducted by civilian scientists in civilian laboratories. As envisioned in *Science: The Endless Frontier*, this type of research would absorb 15 to 30 percent of the foundation's budget.[119]

The armed services consistently supported the establishment of the NSF, insisting though that its military research activities be "coordinated" with their own.[120] Their views, however, mattered very little, for the NSF legislation soon became entangled in several controversies, the most persistent of which was whether or not the foundation would be subject to presidential control. Senator Kilgore's bill called for the President to appoint its director, a provision strongly favored by the Truman administration. Bush's plan, embodied in legislation sponsored by Senator Warren Magnuson, Democrat of Washington, placed responsibility for the appointment of the director, and thus for control of the foundation, in the hands of an independent nine-member commission to be composed of leading scientists and educators serving on a part-time basis. Among the other points of difference between the bills were patent policy and the role of the social sciences. It took six years of wrangling and a presidential veto before the differences could be reconciled and the NSF created.[121]

Bush tried a third line of attack. If independence could not be achieved from without then he would seek it once again from within. In June 1946, he accepted an appointment as chairman of the Joint Research and Development Board (RDB), a defense agency which was established in the postwar reorganization of the military establishment to coordinate the research programs of the services. Soon, though, frustration and exhaustion caught up with him. Within a year, Bush was forced to retire temporarily from public life to rest, being replaced in the RDB post by his long-time colleague, Karl Compton. The RDB, paralyzed by conflicts among the services, came to naught, neither effective in its coordination of military research nor

[119] Detlev W. Bronk, "Science Advice in the White House," *Science* 186 (11 October 1974): 117.

[120] "Services Endorse U.S. Science Set-Up," *New York Times*, 29 May 1946, 25.

[121] The best sources on the National Science Foundation debate are Don K. Price, *Government and Science* (New York: New York University, 1954) and Lawrence McCray, "NSF Legislative History," MIT-ONR History Project, March 1971. See also Don K. Price, "The Deficiencies of the National Science Foundation Bill," *Bulletin of the Atomic Scientists* 3 (1947): 291–94, and Detlev W. Bronk, "The National Science Foundation: Origins, Hopes and Aspirations," *Science* 188 (2 May 1975): 409–14.

viable as a mechanism for bringing civilian influence to bear on military planning.[122]

While all of this was happening, the legislation establishing the Office of Naval Research was quietly making its way through Congress. Admiral Bowen, the representative of a victorious Navy, had no difficulty persuading Congress of the need for an organization to support Navy-related research. At the time, Bush, Compton, and other prominent scientists publicly endorsed the establishment of the new Navy office, describing it as complementary to their own postwar plans.[123] Later, when these plans were unrealized, they would hail ONR as a vital gap filler, an "Office of National Research" when there was none. Still later there were some regrets. James B. Conant, another wartime colleague of Bush, argued in 1971 that if only more pressure had been applied to gain the alternatives that were proposed at the end of the war there would not have been a military research presence in the universities in the 1960s.[124] But without ONR there may not have been the money to support academic research during the years of the greatest growth in science in the United States.

A song said to have been popular among scientists returning to their academic teaching and research responsibilities in 1946 had the title, "Take Away Your Billion Dollars."[125] Some scientists may have wanted to work in an environment totally free of military influence and money. Others apparently wanted an independent role in the planning of military research. None got their wish. Although ONR and the

[122] Don K. Price, *Science and Government*, 144–59.

[123] Hearings on HR 5911 Committee on Naval Affairs, House, 79th Congress, 26 March 1946, 2861–62.

[124] James Bryant Conant, "An Old Man Looks Back, Science and the Federal Government: 1945–1950," *Bulletin of the New York Academy of Medicine* 47, no. 11 (November 1971): 1248–51. Also, "Science Subsidies Traced by Conant," *New York Times*, 9 January 1971, 31 and Interviews. Bush may have been more realistic, arguing back in 1946 that military support was inevitable as well as essential. "Federal Aid Perils Science, Bush Says," *New York Times*, 14 December 1946, 9.

[125] "Take away your billion dollars,
Take away your tainted gold,
You can keep your damn ten billion volts,
My soul will not be sold.

Take away your Army generals;
Their kiss death, I'm sure.
Everything I build is mine, and
Every volt I make is pure." (1946)

Music and lyrics by physicist Arthur Roberts who worked at the M.I.T. Radiation Laboratory. Cited in "A Conversation with Eugene Wigner," *Science* 181 (10 August 1973): 533.

other military research agencies did not quite have billions of dollars to dispense for university-based research, they did become for a period the prime source of support for such research. ONR, at least, was not formed with that mission in mind. Like the scientists themselves, it had to settle for what was possible when its postwar plans were beyond realization.

CHAPTER THREE

The Office of National Research

IN THE YEARS immediately following the Second World War, the Office of Naval Research was the principal federal agency supporting academic science. The NSF was not established until 1950 and did not begin receiving significant appropriations until the Sputnik crisis. Although the Office of Ordnance Research (an Army agency), and the Office of Aerospace Research (an Air Force agency), were patterned after ONR, neither was operating before the early 1950s.[1] The National Institutes of Health, limited in interest to the support of biomedical research, did not collectively match the level of ONR appropriations until the mid 1950s.[2] The Atomic Energy Commission, absorbed by other matters, initially used ONR to manage its support of university-based research. Both the Department of Agriculture and the National Advisory Committee for Aeronautics supported research, but their research interests were narrowly defined and their research clients worked mainly apart from or outside of the universities in specialized research stations.[3]

Senior officials in the Navy Department did not choose such a mis-

[1] The Army established an Office of Ordnance Research in 1951. Being tied to a specific technical branch of the Army (the Ordnance Corps), this office had more limited functions than did either the Air Force or Navy research offices. The Army Research Office was established in 1958. The Air Force Office of Scientific Research (AFOSR), the predecessor organization to the Office of Aerospace Research, was created in 1951 having for one month the dubious title of the Office of Air Research. The Office of Aerospace Research was phased out of existence in 1970 with some of its functions absorbed by the Air Staff's Office of Scientific Research. An excellent history of AFOSR is Nick A. Komons, *Science and the Air Force* (Arlington, Va.: Office of Aerospace Research, 1966). Additional information on the office is contained in Thomas A. Sturm, *The USAF Scientific Advisory Board: Its First Twenty Years, 1944–1964* (Washington, D.C.: Government Printing Office, 1967).

[2] See Stephen P. Strickland, "The Integration of Medical Research and Health Policy," *Science* 173 (17 September 1971): 1093–1103, and his *Politics, Science and Dread Disease: A Short History of United States Medical Research Policy* (Cambridge: Harvard University Press, 1972).

[3] The NACA history is effectively analyzed in Alex Roland, *Model Research: The National Advisory Committee for Aeronautics 1915–1958*, 2 vols. (Washington, D.C.: National Aeronautics and Space Administration, 1985). For descriptions of the agricultural research system, see Jeffrey L. Fox, "USDA Struggles to Reform Its Research," *Science* 225 (21 September 1984): 1376–78, and Vernon W. Ruttan, *Agricultural Research Policy* (Minneapolis: University of Minnesota Press, 1982).

sion for ONR. In fact, involved as these officials were with the demobilization of the fleet and with the battles over unification of the armed services, it would be surprising if they were often conscious that ONR was still in existence. The definition of ONR's mission was literally the responsibility of its own staff. Admiral Bowen's vision for the Office departed with him, for he had been secretive about his ambition to lead the Navy into the nuclear age. What he left behind were the officers and civilians he had recruited to support university-based science in a manner acceptable to scientists and their universities.

As might be expected, the staff found common ground in continuing the program to support academic science. Some believed it important that there be an interim basic research agency until Congress and the President could agree on the structure for the NSF. Others believed it was in the Navy's interest to be on good terms with scientists, particularly those who had gained national prominence because of the success of the atomic bomb project. Still others wanted to preserve a mobilization base for the nation in case of a future conflict. There was no need to rank the motives, for all were well served by the same policies. Different rationales could be cited to different audiences without provoking staff disharmony.[4]

Acting then essentially as the Office of National Research, ONR helped formulate America's postwar science policies, many of which are still with us. It aided the development of academic science, selecting the fields, individuals, and institutions to be supported. It helped devise the contractual forms, the financial arrangements, and the support services for university-based research. It championed the budget for basic science both within and outside government. There was hardly an aspect of the scientific enterprise in America in which ONR was not centrally and constructively involved during the period between the Second World War and the Korean War. Without its sensitive management of what has to be considered limited resources by current standards, the sturdy foundations upon which America's success in science has been built would not have existed.

The Office of Naval Research also helped train a new generation of science administrators and scientists. Its first Chief Scientist, Allan Waterman, became the first Director of the National Science Foundation. Another of its Chief Scientists, Thomas Killian, became the first Chief Scientist of the Office of Army Research. Still another, Em-

---

[4] Some of the rationales are seen in the 1946–1947 speeches of Capt. R. A. Conrad, USN, the Director of Planning at ONR during this period, before such diverse audiences as the Army and Navy Staff College, the Commonwealth Club of San Francisco, the Institutional Research Institute, and a Navy Day ceremony at the University of Illinois. ONR files.

manual Piore, became the Chief Scientist of the International Business Machines Corporation. Among the many other leading science administrators who served in ONR during its first years were F. Joachim Weyl, Mina Rees, Randall Robertson, Urner Liddell, and Roger Revelle. And everywhere today in positions of prominence in American science are the thousands of graduate students ONR supported during its first years.

As the nurturer of postwar science, ONR is held in the highest esteem by scientists. James Killian, the former President of the Massachusetts Institute of Technology and first Special Assistant for Science and Technology to the President of the United States, has said, "The Office of Naval Research represents an extra-ordinary achievement in successful and enlightened government [management of] research."[5] Lee DuBridge, the former President of the California Institute of Technology and also a Special Assistant for Science and Technology to the President, adds, "We will forever owe a debt particularly to the Office of Naval Research for what was done."[6] The extent of that achievement and the extent of that debt are the subjects of this chapter.

BUILDING A RELATIONSHIP WITH ACADEMIC SCIENCE

From a contemporary perspective, it is difficult to imagine the time when there was a need to persuade universities to accept federal support for scientific research. Today, when federal aid for academic science is measured in the billions of dollars, most scientists and university administrators are clamoring for more. But in 1945, when the Navy was preparing to allocate a few tens of millions of dollars to support academic science, there was significant opposition in the universities to a federal subsidy for research no matter how modest the potential offer. Some then feared that federal support for science would lead to centralized political control of science. Others worried about the stability of the federal effort, subject as it always would be to the uncertain politics of the budgetary process.

The fear of political control of science stemmed from two quite different concerns. On the one hand there was the concern that federal subsidies were a potential mechanism for restricting the freedom of

---

[5] Cited in W. D. Brinckloe, "Research Navy," *American Society of Naval Engineers Journal* 71 (February 1959): 96.

[6] Ibid. See also John E. Pfeiffer, "The Office of Naval Research," *Scientific American* 179 (February 1949): 11–15; *Science and Public Policy*, Report to the President [Steelman Report] (Washington, D.C.: Government Printing Office, 1947), I:55; Lee A. DuBridge, "Science and National Security," *Science* 120 (31 December 1954): 1081–85.

inquiry. Fresh in mind were the wartime security measures imposed in the United States which had limited communications among American scientists and the fate of German science under the Nazis, the triumph of ideology and ignorance. On the other hand there was the concern that federal subsidies could alter the structure of the scientific community. The competition for private resources had always been intense in American science. Those who had been most successful in this competition could not be certain that the addition of another source of support would not be disruptive, creating new allocation standards and new status rankings. If the federal government was not likely ever to promote totalitarian interests, it might easily come to promote equalitarian interests.[7]

University administrators also had more prosaic fears. The rapid wartime mobilization of science had been followed at the completion of the war by an equally rapid demobilization. Would the postwar program to support science be any more stable even if more modest in scale? Government policy, the more sophisticated among them knew, was dependent on the durability of interest coalitions. Would the coalition supporting the allocation of funds for academic science be long lasting?

The M.I.T. Corporation, that university's board of trustees, at least, did not think so. Karl Compton, the President of M.I.T., reported to an early meeting of the Naval Research Advisory Committee (a body created by the statute which established ONR) that the Corporation believed federal funds were the most unreliable source of support available to M.I.T. The Corporation, he confided, felt that alumni and business support should be preferred over federal support in building M.I.T.'s future.[8]

To be sure, these fears presented no great obstacle to the initiation of the Navy's academic research program. Then as now, the allure of federal money was difficult to resist. The competitiveness of the American university system alone guaranteed that there would be willing takers for the research support the Navy had to offer. If M.I.T. or Harvard did not want the money, Berkeley or Chicago might. And because university faculty were mobile, the universities that considered refusing the support knew that they would risk the loss of the most

---

[7] Alexander G. Ruthven, "Leadership or Regimentation in Higher Education," *Educational Record* 18, no. 3 (July 1937): 345–53; Frank B. Jewett, "The Future of Scientific Research in the Postwar World," in John C. Burnham, ed., *Science in America* (New York: Neale Watson, 1971, 398–413); James B. Conant, *My Several Lives: Memoires of a Social Inventor* (New York: Harper & Row, 1970), 117.

[8] Transcript of the second meeting of the Naval Research Advisory Committee, Washington, D.C., 15 January 1947, 47.

valued members of their staffs to a rival if in fact they did refuse the Navy's offer.

Nevertheless, Admiral Bowen was sensitive to both the vulnerability of the universities and their fears. In the fall of 1945, before ONR was formally established, he dispatched Capt. Robert Conrad, the officer who was to be responsible for the initial planning of the academic research program, to Harvard, M.I.T., Chicago, California Institute of Technology, and Berkeley to gain the acceptance of university officials at these institutions for the new venture.[9] Conrad promised that the program would be designed so as to neither threaten the integrity nor the stability of the universities. The Navy would pay the full costs of the research. Only the best projects would be supported; most would be initiated by the scientists themselves. Very little of the work would have to be classified. All fields of science would be included. And there would be continuity in the support.[10] With each of these major research universities fearing that they would be at a competitive disadvantage if they did not accept the Navy's money, all agreed to participate in the new program. Once the leading research institutions had signed on, most other universities quickly followed their example.[11] And because ONR strove to keep the promises that Conrad had made, none had any initial regrets that they agreed to accept the money.

TEMPERING THE RESEARCH CONTRACT

Two prime instruments by which government has supported academic research initially represented contrasting conceptions of governmental purpose. One, the research contract, implied a quid pro quo relationship in which government was purchasing a service or a piece of equipment for an agreed-upon price from a supplier, in this case a university-based scientist or research group. The other, the research grant, implied a gift relationship in which government encouraged an activity in which the scientist or group was already involved or wished to undertake. When a contract was used, the awarding agency was obligated to act like a buyer in a market, advertising its intention to purchase and seeking the best possible terms and a tangible benefit for government from the transaction. In contrast, when a grant was

---

[9] Memorandum to Chief of ORI from Capt. R. D. Conrad and Lt. Comdr. B. S. Old: Report of Visit to University of Chicago, University of California, and the California Institute of Technology (n.d.). Also J. H. Faull Jr., "Two Autobiographies: J. B. Conant and Vannevar Bush," *Newsletter: ONR Boston*, 31 December 1970, 18.

[10] Interview documents.

[11] Interview documents. Also, "Navy Begins Financing Research in Colleges Training Scientists," *New York Times* 27 January 1946, 1.

used, the awarding agency was required merely to be a disinterested benefactor seeking to identify and support a meritorious beneficiary.[12]

Not surprisingly, because a grant implied merit and imposed few, if any, reciprocal obligations on recipients, scientists preferred to receive their support in the form of grants rather than contracts.[13] University administrators, however, had at least an equally strong preference for the award of contracts. If research support were considered to be a gift, government was not likely to pay the full cost of the research supported. Just as was the practice of the private foundations, government could and did insist that its grant recipients participate in the financing of the work to be supported, as they were seen as both the initiators of the research and its prime beneficiaries. If research support were considered to be not a gift but a purchase, government could be charged and would pay all the costs involved in the work it requested, including what were termed overheads or indirect costs, a marvelously effective way to finance growth in the universities.

The legislation that established ONR authorized the use of contracts for research, but not grants.[14] Because the Navy had traditionally used contracts in its procurement activities, no attempt was made to secure granting authority for ONR. To have done so, given the prevailing attitudes, would have been to signal an entirely different, and perhaps unacceptable, purpose for ONR than the one acknowledged. It was not until 1959, in the aftermath of the Sputnik crisis, that defense research agencies (including ONR) were authorized to award grants.[15]

With only contractual authority, ONR had to satisfy the conflicting desires of its two academic constituencies: the scientists who hoped for the freedom of grants, and the university administrators who needed contracts in order to recover for their institutions the full costs of government-financed research. The way out of the organizational problem, the ONR staff discovered, was to blur the operational distinction between contracts and grants by making the Navy's research contracts as unburdensome to the researcher as any grant and as finan-

[12] See Committee on Institutional Research Policy, *Sponsored Research Policy of Colleges and Universities* (Washington, D.C.: American Council on Education, 1954); *Federal Bar Journal* 17, no. 3 (July–September 1957); *Administration of Research and Development Grants*, Report of the Select Committee on Government Research of the House, 88th Congress, 2d Session (Washington, D.C.: Government Printing Office, 1964) and especially, Charles V. Kidd, *American Universities and Federal Research* (Cambridge: Harvard University Press, 1959).

[13] Steelman Report, 5:60.

[14] Ibid., 2:11.

[15] "Research Grants to be Used by Defense Department," *Science* 130 (16 October 1959): 969.

cially rewarding to the researcher's university as any contract a government agency could award.

The precedent set by OSRD helped the situation considerably. Much to the displeasure of the financial managers in the Bureau of the Budget and the General Accounting Office, Vannevar Bush, by direct appeal to the House Appropriations Committee, had won the right for OSRD and other government agencies to pay universities full costs on research contracts during the Second World War.[16] The agency had also breached the requirement that the availability of contracts be formally advertised, thus permitting the negotiation of research awards with selected universities just as was occurring in military procurement contracts with industry.[17] But it was ONR's own administrative flexibility that made the formal difference between contracts and grants inconsequential.

If scientists held the initiative in the selection of research problems to be investigated in the award of grants, so too would they hold the initiative in the selection of research problems in the award of ONR contracts. Senior ONR officials determined the division of the agency's research support among the scientific disciplines, but the choice of specific projects to be supported was the responsibility of civilian program officers who were themselves qualified scientists in the disciplines in which the awards were to be made. University-based scientists were encouraged to submit as proposals the projects they would like to pursue, regardless of the project's potential value to the Navy.[18] The program officers' decisions as to which projects to support among those submitted were based on their own judgments, sometimes supplemented by consultations with the discipline's elder statesmen, as to which of the disciplines' problems seemed the most important to solve and which investigators seemed most likely to solve them.[19]

[16] Vannevar Bush, *Pieces of the Action*, 1008; Interview document.

[17] The authority of OSRD and ONR to waive advertising came from the First War Powers Act. Subsequently, the waiver authority was contained in the Armed Forces Procurement Act of 1947. Edwin P. Bledsoe and Harry I. Ravitz, "The Evolution of Research and Development as a Procurement Function of the Federal Government," *Federal Bar Review* 17, no. 3 (July–September 1957): 205. Bledsoe served as Deputy Director of ONR's Contract Division and Ravitz as Head of the Contract Administration and Distribution Branch at the time their article was published. Both were key figures in the design of the ONR research contract.

[18] Interview documents.

[19] Interview documents. For a discussion of how an early ONR program was managed, see Lawson M. McKenzie, "A View of Science and Technology Spring from Cryogenics," *Naval Research Reviews*, January 1966, 2–13, and Lawson M. McKenzie, "After Six Years—A Study of the Impact of the Physics Branch Program," *Science* 118 (28 August 1953): 227–32; Frederick Seitz, "The Governmental Science Administrator," *Physics Today*, August 1961, 36–38.

There was no need, it was thought, to rank the proposals other than on the basis of their perceived scientific merit in order to serve the Navy's interest in the awards, as the belief then within ONR was that all basic research had potential naval applications.[20] And because the program officers were very much part of the disciplines they served, most often recruited from them and expecting to return to them, it was the scientific community itself that determined the merit ranking and the contract awards.

If research grants had educational benefits, so too would ONR research contracts. Investigators were encouraged to include support for graduate students in the budgets of their ONR projects. With the war over, universities were anxious to expand their advance training programs in the sciences which had been curtailed by the draft of students and the mobilization of faculty. ONR was prepared not only to finance current research activities, but also to help build the research base for the future. It actively sought, not merely acquiesced to, the participation of graduate assistants in its contract research.[21]

And if every other agency's research contracts were considered especially onerous by academics because of their demanding reporting requirements, ONR's would not be so burdened. The ONR staff acted as if it were the scientific findings that were important in research and not the adherence to standard administrative procedures. Rather than insisting, as did other agencies, that university research investigators file time-consuming monthly or quarterly activity statements on their contracts, ONR contract monitors usually required only that an annual report be submitted. Because so few ONR contracts were classified (only fifteen of seven hundred contracts involved secret research in 1948, for example [22]), almost all ONR contractors could publish their

[20] "Almost 50 percent of the ONR contracts are in basic research with no 'applied' strings attached, for the Navy is fully aware that a sound basic research policy is the foundation of later developments on the applied side. Most of our projects come in as research proposals originating with the individual." Letter to the Editor, *Washington Post*, 12 November 1950 by Rear Adm. T. A. Solberg, Chief of Naval Research. Reprint, *American Psychologist* 6, no. 3 (March 1951): 94. Interview documents.

[21] James R. Killian, Jr., "Government Research in Educational Institutions: Its Benefits and Hazards," a paper presented at the Naval Research Conference, 18–19 November 1947, Washington, D.C. Conference Proceedings (Washington, D.C.: Navy Industrial Association, 1948), 91. Note that twenty-five hundred graduate students were working with support from ONR contracts in 1947. Note also Benjamin Fine, "Navy and Colleges Cooperate in Largest Peacetime Program of Scientific Research," *New York Times*, 10 February 1946, 9. Joan H. Criswell, "Support of Graduate Students in Social Psychology by the Office of Naval Research," *American Psychologist*, 9, no. 4 (April 1954): 148–50; and Mina Rees, "The Saga of American Universities: The Role of Science," *Science*, 179 (5 January 1973): 21.

[22] Transcript of the sixth meeting of the Naval Research Advisory Committee, Washington, D.C., 20 January 1948, 3.

research results in the open literature, thus gaining the benefit of professional recognition of their work. In fact, ONR often accepted the submission of journal articles or book manuscripts as proof of project completions in place of extensive final reports.[23]

It is doubtful scientists had ever encountered a more accommodating patron. ONR willingly sponsored conferences on topics of interest to its contractors and was generous in the provision of travel support in order to facilitate communication among investigators. It assisted scientists in obtaining surplus military equipment for their laboratories and in gaining access to naval test ranges, ships, and aircraft for the conduct of field experiments. It gave scientists control over the direction of its research program, believed in open research, and even liked graduate students. Not surprisingly, many scientists soon came to prefer ONR's contracts to any other agency's grants.

But ONR courted the goodwill of university administrators as well as that of scientists. While ONR was ready to concede the intellectual direction of research to scientists, it always deferred to the preferences of university administrators when it came to the financial management of research. Again, it was ONR's administrative flexibility that was crucial to its success in building mutually supportive relations with university managements.

This flexibility was demonstrated, for example, in its handling of the problem of faculty income increases due to ONR-sponsored research. Although ONR's contract research program initially involved expenditures of only about $20 million, the agency quickly realized that its research support could disrupt the entire pattern of university compensation if faculty members were permitted to significantly increase their income because of contract support. A few universities were willing to allow this to happen, and indeed had made provision for faculty consulting on sponsored research or the payment of overtime to reward faculty with contracts, but most were not. Not wishing to be the cause of precipitous change, ONR sought to temper the impact of its largess. Initially, it attempted to avoid providing any salary support for senior faculty members. Later, it permitted term time release and summer supplements to be drawn from contracts. But there was no consistent policy other than to be certain that ONR did not dictate the salary schedules for the universities. The desire was to follow, not to lead.[24]

The deference of the ONR toward university administrators was

[23] Interview documents.

[24] Interview documents. Transcript of the fourth meeting of the Naval Research Advisory Committee, Washington, D.C., May 1947, 114 and transcript of the twelfth meeting of the Naval Research Advisory Committee, Washington, D.C., 22 May 1950, 59ff, 129ff.

clearly shown in issues relating to overhead costs. Early in its history, several of the scientists on its staff strongly advocated requiring universities to cost-share on contracts in order to increase the amount of dollars available for research. If there were cost-sharing, more projects could be supported for the same budget. But senior ONR officials, not wishing to renege on the promise made to university administrators that ONR would provide full overhead reimbursement, rejected the advice.[25] A few years later, when one or two universities with apparently small expectations of major contract work offered to cost-share on a voluntary basis, ONR refused after soliciting the advice of the presidents of several of its major contracting institutions. There was no reason, they suggested, to set avoidable precedents. And ONR never found the need to seek recovery of overhead reimbursements it paid to the state universities that collected contract overheads and state legislative appropriations for the same indirect costs.[26]

The promise to maintain stable support was also not forgotten. In 1948, as part of general reductions imposed upon the military, ONR's university research appropriation was cut in half from $22 million to $11 million.[27] Rather than reduce its support of university research, however, ONR drew from the reserve set aside for projects continuing over several years in order to maintain the dollar flow to the universities and permit the award of new projects.[28] It then pressed for a supplemental appropriation to extend the average term of its contracts which had had to be reduced because of the expenditure of the reserve.[29] The unfunded years of approved projects were considered "moral obligations," if not legal obligations, in subsequent budget presentations.[30] Congressional appropriations over the next few years were sufficient to keep the university research program growing, but were not enough to add to the average length of the contracts they supported. With the surge in military appropriations that accompanied the Korean War, a supplement of $20 million was obtained on

[25] Interview document. Transcript of the twentieth meeting of the Naval Research Advisory Committee, Washington, D.C., 16 June 1953.

[26] Interview document.

[27] Hearings on H.R. 3993 before Naval Affairs Subcommittee, House Appropriations Committee, 80th Congress, 1st Session, 21 February 1947, 1548–51.

[28] Transcript of the sixth meeting of the Naval Research Advisory Committee, Washington, D.C., 20 January 1948, 67–78.

[29] Transcripts of the tenth meeting of the Naval Research Advisory Committee, Washington, D.C., 19 September 1949, 8ff and eleventh meeting of the Naval Research Advisory Committee, Washington, D.C., 30 January 1950, 3ff.

[30] Interviews. For a discussion of recent problems in maintaining award stability, see Barbara J. Culliton, "NIH Pays a Price for 'Stability,' " *Science* 221 (15 July 1983): 243–44.

the pledge that these funds would be used only for contract "longevity."[31] At the next reduction in its budget, however, ONR again dipped into the longevity reserve for a compensating amount.[32] Risky budget tactics, perhaps, but given their purpose, actions certain to be appreciated in the universities.

When ONR, along with other defense agencies, finally did gain authority in 1959 to award research grants as well as research contracts, there was little reason to use it. By then, both scientists and university administrators were content with the ONR research contract.[33]

THE ORGANIZATIONAL DESIGN

Public Law 588, which created ONR, not only gave it equal status with the Navy's materiel bureaus, but also gave it independence from the fleet and the growing authority of the Chief of Naval Operations. The Navy in 1946, as it had been for nearly a hundred years, was a bilinear organization. That is, there were two parts to the Navy: the fleet itself, and the shore-based materiel support establishment, each reporting separately to the Secretary. The Chief of Naval Operations emerged in the immediate postwar years as the dominant officer within the Navy, having control over the fleet and most planning activities. He did not, however, have control over ONR, which reported directly to the Secretary.

This formal independence from the uniformed Navy was both an advantage and a disadvantage. It was an advantage in that ONR was neither absorbed by the routine demands of the fleet nor subject to the reorganization whims of senior naval officers. But ONR was also somewhat isolated from the main interests of the Navy—the current and future health of naval forces. While ONR in its formative years was free to focus its energy on the support of university research, it could not shake a lingering fear that most of the rest of the Navy, obviously indifferent to its existence, could easily turn hostile.

To be sure, ONR as a naval agency was always headed by a naval officer; the law required it. Rear Adm. Paul F. Lee, a ship design engineer, replaced Admiral Bowen as Chief of Naval Research in 1946. Three years later, Admiral Lee was in turn replaced by Rear Adm. T. A. Solberg, an officer with experience in the Navy's nuclear weapons program. But despite some effort on their part and that of others in

---

[31] Letter from Alan T. Waterman to members of the Naval Research Advisory Committee, 12 February 1951.

[32] Transcript of the twenty-first meeting of the Naval Research Advisory Committee, Washington, D.C., 17 November 1953.

[33] Interviews.

the Office, it was difficult to attract first-rate officers to ONR. Few naval officers had an interest in science; even fewer wished to become involved in an organization so apparently divorced from the mainstream of naval activities.

Recruiting scientists for the ONR staff was itself a difficult task, but for a different reason. Government salaries at the time were low, even relative to those offered in the universities. Legislation in 1947 establishing a special pay classification for a limited number of professional and scientific personnel in each of the services (Public Law 313) helped ease the problem, as did the view among scientists that a tour in ONR was a potential stepping-stone for responsible research management or administrative positions within the universities. Some senior scientists even argued that there should be a tradition to serve in ONR as one of the obligations of a scientific career.[34]

Capt. Robert Conrad, who initially designed ONR's university contract program, might have been able to integrate ONR more closely with the rest of the Navy had he been able to remain longer in his position. A forceful officer with strong beliefs about the necessity of tying research to naval applications, Conrad had been with the Office of the Coordinator of Research and Development during the Second World War and had served as a key deputy to Admiral Bowen during the establishment of both the Office of Research and Invention, and ONR. Conrad, though, was required to retire early from active duty when he was discovered to have terminal leukemia in 1947.[35]

The basic design Captain Conrad conceived for the university contracts program was also bilinear in form (see Fig. 3-1). A Science Branch subdivided into disciplinary sections such as Physics, Electronics, Chemistry, Mathematics, Physiology, Psychology, and Biochemistry would interact with university scientists, survey research opportunities, and award contracts. A Program Branch with sections labeled Armaments, Amphibious Warfare, and Power would identify naval problems where scientific developments might be applied. With few exceptions, scientists staffed and led the science branch sections, while naval officers were in charge of the program branch sections. The independence of the biomedical disciplines from the physical sciences was soon recognized in the creation of a parallel Medical Sciences Branch headed by a naval physician, but the basic division between scientists concerned with university research and naval officers

---

[34] Transcript of the fifth meeting of the Naval Research Advisory Committee, Washington, D.C., September 1947.

[35] Interviews. "Robert Dexter Conrad (1905–1949)," *Nucleonics* 5, no. 3 (September 1949): 82–83.

FIGURE 3–1. Organizational Chart, Office of Naval Research, October 1946.

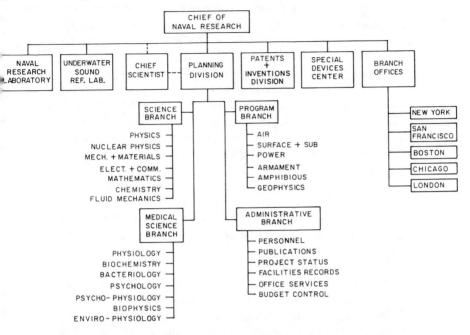

concerned with the requirements of naval warfare persisted. The linkage between the two was the task Conrad had hoped to accomplish.[36]

After Conrad's retirement, control of the Office fell away from the military to the civilian staff members. Alan T. Waterman, the Yale physicist who served as ONR's first Chief Scientist, is often credited with being the organization's key figure during its formative years. His reputation as the crucial builder of ONR, however, appears somewhat exaggerated, no doubt embellished by the fact that he moved on in 1951 to become the first Director of another important government patron of research, the NSF. Waterman's role seems on examination to have been confined mainly to external relations, conferences with university presidents and presentations before Congress.[37] Instead, the direction ONR took was set by younger scientists on the staff, especially Emmanual Piore, Mina Rees, and Randal Robertson who were stimulated by their daily contact with university-based scientists. Together they saw opportunities for advances in basic knowledge created by the technical achievements of the war. European science, the fountainhead of past discoveries, was devastated. American science was ready to seize the leadership in research progress, but was in

[36] Capt. Robert Conrad, diary, 27 May 1956, ONR files.
[37] Interview document.

need of support. ONR had some money, but no clear mission. Because the military had not yet absorbed the new technology produced by the war, it was obvious to ONR's scientific staff that the organization's attention had to focus on building up the resource base for basic research in American universities.[38]

The agency's initial budget priorities reflected this judgment. Less than 10 percent of the more than $86 million appropriated to the ONR contract research program in the period fiscal year 1946 through fiscal year 1950 was spent to support naval science applications.[39] The bulk of the funds, augmented by additional millions transferred to ONR by the Atomic Energy Commission, were allocated to buy equipment and support research and training in the physical sciences and, to a lesser extent, in the biomedical sciences. ONR's contract research, concentrated in the universities, amounted to more than 15 percent of the total Navy R&D expenditures during the years before the Korean War, but given its distribution among fields, it was serving national as well as naval purposes.[40]

To assist the Washington-based staff in the management of widely scattered contracts, branch offices were established in San Francisco, Los Angeles (later moved to Pasadena), Chicago, New York, and Boston.[41] Most had their origin in the liaison offices the Navy had created near major university research facilities during the war. Like the Washington central office, the branches were staffed by a mixture of naval and civilian personnel. The branches helped locate potential contractors, monitored contract performance, aided in the processing of the inevitable paper work, and sought out surplus equipment for universities. Control of the scientific program, however, remained firmly in Washington.[42]

The most famous branch office did little contract-related work and was not even located in the United States. ONR London was the direct descendant of the wartime OSRD London office. Its task was to act as a window on European science, reporting on the latest developments and to assist visiting American scientists to make contact with their

[38] Interviews.

[39] Calculated from ONR Funding History table Code 52, 4 November 1970; "The Chief of Naval Research Reports to the Naval Research Advisory Committee," 27 April 1961, 17; letter from J. Jayne, ONR Budget Office to W. Weaver, Chairman NRAC, 5 May 1947.

[40] Interviews.

[41] Conrad considered expanding the number of branch office sites to include Seattle, St. Louis, Austin, Atlanta, Pittsburgh, and Los Angeles. Conrad diary, 25 July 1946, ONR files.

[42] As one branch office secretary put it succinctly: "What are you interviewing here for? Everything important happens in Washington." Interview document. See also Conrad diary, 3 April 1946, ONR files and Interview documents.

colleagues in Europe, a courtesy most deeply appreciated, especially in the chaotic first years after the war.[43]

Besides the contract research program, ONR had other responsibilities not all of which it performed with enthusiasm. It supervised the Navy Patent Program, complete with field sites at naval installations and a portfolio of several thousand patents. It sponsored the Naval Research Reserve, which in 1950 had over five thousand members, and which subsequently produced at least two Rear Admiral-scientists. It financed the advanced science training of naval officers. And it had the task of coordinating research within the Navy.

As a legacy of Admiral Bowen's tenure as Chief of Naval Research, ONR was also officially responsible for two major laboratories. The largest, the Washington-based Naval Research Laboratory (NRL), with over one thousand professional staff members and a budget approaching that of the contract research program, was a reluctant ward after Bowen's departure. Because ONR contracted with universities for research similar to that being conducted at the laboratory, and because NRL itself maintained a contracts program, the relationship between the organizations was always somewhat strained. As long, though, as ONR strove to keep NRL fully funded, the laboratory refrained from criticizing ONR's university research effort or offering itself as an alternative program manager.

The Special Devices Center, the other major laboratory under ONR nominal control, was the child of Admiral de Florez, the flamboyant aviator-inventor and close friend of Admiral Bowen. The Center, originally located in Washington, D.C., moved to a mansion at Sandy Point, Long Island at the end of the war where it supposedly continued to work in developing training equipment. Because Admiral de Florez became the first director of technical research at the Central Intelligence Agency about the time his Special Devices Center moved to plusher quarters, what it did, if anything, beyond producing simulators for the Navy, is not clear. In any event, ONR benefited from the Special Devices connection for it received not only Admiral de Florez's counsel as a member of its Naval Research Advisory Committee, but also the transfer of over $30 million from the Bureau of Aeronautics, the Special Devices Center's original home agency.[44]

[43] Plans for a similar office in Latin America never materialized. The Tokyo office was finally opened in 1975. Interview documents. On the Tokyo office, letter from Dr. Walter H. Bratain to Dr. James W. McKae, 8 July 1959, NRAC files and report of ONR Far East Branch Office Survey Team, 24 July 1959, ONR files. Note also John Walsh, "ONR London: Two Decades of Scientific Quid Pro Quo," *Science* 159 (4 November 1959): 623–25. As part of its activities, the London office publishes *European Scientific Notes*, an excellent source of news on science and science policy in Europe.

[44] Admiral de Florez believed that it was this money that saved ONR from an early

JURISDICTIONAL RELATIONS

Although there was general agreement within ONR that its preferred mission was the support of basic research, there was disagreement on whether that mission was to be permanent or temporary. For some of the staff, ONR was thought to be only a gap filler until the NSF could be established, the legislation for which was caught in conflict between Congress and the President, not to be resolved until 1950. For others, ONR was thought to be a permanent alternative source of support for science. Yet, from its very beginning, steps were taken to assure ONR's survival no matter the fate of the effort to create the NSF.

The legislation establishing ONR contained a section stating that ONR was to "coordinate the research activities of the Department of the Navy," which, given the structure of the Navy, meant the research activities of the materiel bureaus. ONR's leaders, however, made no serious attempt to implement the coordination mandate. They knew that the words "to coordinate" had been hastily substituted for the words "to control" in the draft of the bill when the bureaus threatened to oppose its passage on the grounds that ONR was being given power to direct their research planning and management activities.[45] So as not to provide the bureaus with the slightest concern that the newly created agency coveted any of the bureaus' functions, they avoided actions that would imply ONR was criticizing or reviewing the bureaus' research programs. Instead, they made ONR seem the willing servant of the bureaus, offering to perform administrative and contracting tasks at the bureaus' direction.[46] The agency sought no enemies within the Navy.

When it came to relations with the other services, however, ONR was prepared to take a more imperialistic stance. Several times during the late 1940s, ONR approached the Research and Development Board (RDB), the agency charged with the harmonization of research planning within the Department of Defense, with proposals to act as the central defense planning agent for basic research. The RDB was being pressed by the Bureau of the Budget to eliminate duplication in defense research, but it had been unable to develop an effective mech-

---

demise. Transcript of the ninth meeting of the Naval Research Advisory Committee, Washington, D.C., June 1949, 75. The history of the Navy's work in training devices is discussed in Charlotte A. Hambley, "Genuine Fakes," *Sea Power*, March 1987, 34–43.

[45] Interviews.

[46] ONR's position contrasts sharply with that of the Air Force's Office of Scientific Research which continually struggled with various Air Force laboratories and commands for control over external research contracting authority. Robert Sigethy, *The Air Force Organization for Basic Research 1945–1970: A Study in Change*, unpublished doctoral dissertation, American University, 1980.

anism to control the proliferation of basic research projects. Because neither the Army nor the Air Force had then created an organization similar to ONR, the proposals had a selfless appearance. Yet, had they been accepted—and only the territorial instincts of the fledgling Air Force prevented their acceptance—ONR would have gained precisely the veto power over Army and Air Force research that it would not consider seeking within the Navy. The proposals were not innocently offered, regardless of their outward appearance.[47]

Relations between ONR and civilian agencies were quite variable. Those with the Atomic Energy Commission (AEC), for example, deteriorated over time. ONR was happy to manage the AEC's basic research program in nuclear physics while the AEC was getting organized during the late 1940s. But when the AEC was ready with congressional urging to take over the program, ONR made it clear that it had no intention of conceding the field entirely. ONR wanted to share the contracts with university-based physicists, many of whom preferred their ties with the Office.[48] The agency had excellent relations with the National Institutes of Health, primarily because it limited its own support of the biomedical sciences. The ONR staff members specializing in biology, dissatisfied with the allocations they received, quickly shifted over to the NSF when it was established.[49] The relations with the State Department were always contentious. The State Department sought jurisdiction over the ONR London Office which ONR refused to relinquish. State, it was thought, would be certain to bungle the job of maintaining the ties with European science if it had the responsibility, and it was extremely doubtful that the State Department could recruit scientists to do the work. There was no point, it seemed, in giving the department any meaningful role in the foreign relations of science.[50]

Crucial to ONR's survival was the delay in establishing the NSF. If NSF had been an operating agency in 1947 or 1948 rather than in 1951, ONR would have been forced to give up much of its academic research program.[51] Although ONR was always on record as favoring the early creation of NSF, advocates of NSF (including Vannevar Bush), suspected that ONR's support was less than nurturing.[52] As ONR be-

[47] Transcript of the tenth meeting of the Naval Research Advisory Committee, 25 September 1949.

[48] Ibid.

[49] Milton Lomask, *A Minor Miracle: An Informal History of the National Science Foundation* (Washington, D.C.: National Science Foundation, 1976), 74–77.

[50] Transcript of NRAC meeting, 19 September 1949.

[51] James Bryant Conant, "An Old Man Looks Back, Science and the Federal Government: 1946–1950," *Bulletin of the New York Academy of Medicine* 47, no. 11 (November 1971): 1248–51.

[52] Alan T. Waterman, diary notes, conference with Dr. Bush, 11 March 1947.

came more secure in its relations with the scientific community, it became bolder in its efforts to hobble, if not block, NSF. So successful were these efforts, that ONR had little to worry about when NSF finally was established. NSF seemed, then, according to one high ranking ONR official, "almost part of the family."[53]

The agency's attitude toward NSF was set early, statements calling for its creation or promising the transfer of projects notwithstanding.[54] As the final report of the Office of Research and Invention expressed it in 1947:

> Only by having contracts for basic research in each field of science can the Navy be certain of a continued appreciation of scientists and their methods and can the Navy organization continue to receive the stimulation of independent creative minds. Moreover, it is desirable for scientists to have a continued awareness of their responsibility to the national welfare and security, which association with the Navy brings. . . . The establishment of an NSF . . . does not relieve the Navy Department of its responsibility to prepare the Navy for war.[55]

An agency like the Office of Research and Invention, e.g., ONR, according to the report, would be needed if the Navy was to get the benefits of a civilian research agency.[56]

To be certain that NSF itself would not usurp this role, ONR sought to have the NSF bill changed so that there would not be a separate military division created within NSF. A separate division, of course, would be a direct rival to ONR. What ONR proposed instead was a diffusion of military interests within NSF; something that scientists would most likely find unacceptable and would seek to dismantle once NSF was operating.[57]

As the likelihood that there would be an NSF increased, strategies for limiting its impact were discussed within ONR and with representatives of the National Institutes of Health (NIH), an agency also po-

[53] Comments of Emmanual Piore, transcript of the sixteenth meeting of the Naval Research Advisory Committee, 22 October 1951, 74.

[54] Philip N. Powers, "A National Science Foundation?" *Science* 104, (27 December 1946): 614–19. Powers was then an ONR staff member.

[55] Office of Research and Invention, *Annual Report FY 1946* (Washington, D.C.: Department of the Navy, 1947), 114–15.

[56] ORI, *Annual Report FY 1946*, 114. This view was also repeated in a letter from Admiral Lee, Chief of Naval Research, to Congressman Charles A. Wolverton, 26 May 1947.

[57] Transcript of the third meeting of the Naval Research Advisory Committee, 26 March 1947, 54–55. As reported at the fourth meeting of the Naval Research Advisory Committee, May 1947. The Senate sponsors of the bill rejected the proposal, though, noting that the Navy and the Army had asked for the division in the first place.

tentially threatened by NSF's creation.[58] NSF, it was suggested, should be encouraged to concentrate on the institutional support of science and on the development of programs for increasing scientific manpower while ONR (and NIH) would take the lead in supporting project research.[59] Few research projects would be transferred to NSF and those that were, would be projects that ONR did not intend to fund.[60] "Give them projects we had on the shelf," was one way it was put in internal discussions.[61]

The Office of Naval Research was not to be unhelpful to the new agency. On the contrary, it was thought best to offer assistance and administrative support in order to avoid a charge of obstructionism.[62] But along with the aid was the clear message that the Navy had to continue its independent support of academic science. There were problems in science of special interest to the Navy, requiring it to remain in contact with outstanding scientists. Just before the NSF bill was enacted into law, a resolution was passed by the Naval Research Advisory Committee calling for the pluralistic support of science.[63]

Consideration was also given to working with the Bureau of the Budget (BOB), the agency that would be responsible for organizing the NSF, setting up its accounts, allocating personnel, and the like. The fear was that BOB would independently force a major transfer of personnel and projects to NSF. Fortunately, the BOB staff person assigned to the task, Spencer Platt, was a friend of an ONR official, who reported that Platt did not know anything about research. The suggestion was

[58] Transcripts of the ninth and eleventh meetings of the Naval Research Advisory Committee. Interview documents. NIH was determined to limit NSF's interest to traditional biology while retaining for itself all biomedical and medical research. Swain discusses NIH-NSF relations in his article on the growth of NIH, Donald C. Swain, "The Rise of a Research Empire: NIH, 1930–1950," *Science* 138 (14 December 1962): 1236.

[59] Comments of Dr. McCann, transcript of the ninth meeting of the Naval Research Advisory Committee, June 1949, 77. NSF's early emphasis was on science manpower programs, especially graduate fellowships. Note for background M. H. Trytten, "The New Science Foundation," *Scientific American* 183, no. 1 (July 1950): 11–15.

[60] Interview document; Kent C. Raymond and Thomas M. Smith, "Project Whirlwind: A Case Study in Contemporary Technology," processed Smithsonian Institution, 1969, chap. 6, p. 17; Thomas J. Killian, "The National Science Foundation and Research," Proceedings of the fourth Annual Conference on the Administration of Research (Ann Arbor, Engineering Research Institute, University of Michigan, 1951), 70–74.

[61] Transcript of the ninth meeting of the Naval Research Advisory Committee, June 1949, 85.

[62] Comments of Admiral de Florez, transcript of the fourth meeting of the Naval Research Advisory Committee, June 1947, 90. ONR did provide administrative support for NSF when it was established. Minutes of the fifth meeting of the National Science Board, 6 April 1951.

[63] Transcript of the eleventh meeting of the Naval Research Advisory Committee, 30 January 1950, 129ff.

made to discuss NSF arrangements with Platt and give him a plan for its organization. "Tell him we studied this."[64]

As it turned out, ONR's influence on NSF could hardly have been greater; seven of the foundation's top ten officers had either an ONR or Navy connection, among them Allan Waterman, ONR's Chief Scientist, appointed as the NSF's first director.[65] Waterman, of course, had been part of the discussions held at ONR on ways to stifle NSF. One can imagine his embarrassment when he approached ONR as he had to for projects to be transferred to NSF. Yes, there were a few that should be of great interest to the NSF, he was told.[66]

The National Science Foundation had been expected to be supporting $250 million worth of research by 1957 when the bill calling for its establishment was being debated in 1947.[67] Its first real appropriation request in 1952 was for $14 million. But Congress was unenthusiastic about the new agency and cut the request to $3.5 million.[68] NSF would not gain significant appropriations until it began, after Sputnik, to use Cold War justifications for its requests.[69] In the meantime, ONR continued to function as the nation's prime source of support for basic research.

[64] Transcript of the ninth meeting of the Naval Research Advisory Committee, June 1949, 86–88.

[65] See list of NSF officials in Dwight Gray, "NSF One Year Later . . ." *Physics Today* 5, no. 7 (July 1952): 13. NSF continued to draw on ONR for its senior officials through the early 1970s, though at a diminishing rate and not always by its own choice. Note: "White House Nominates Four to Long-Unfilled Posts," *Science* 168 (3 April 1970): 101. Waterman was the seventh choice of the NSF Board for Director, not the first or even the third as some sources have stated. Minutes of the second meeting of the National Science Board, 3 January 1951.

[66] Draft notes of a conference with representatives of the Office of Naval Research, 22 January 1952. Waterman papers. See also Alan T. Waterman, "Research for National Defense," *Bulletin of the Atomic Scientists*, 9, no. 2 (March 1953), especially p. 37.

[67] Steelman Report, 31.

[68] The budgets for the NSF's first years are examined in Dael Wolfle, "National Science Foundation: The First Six Years," *Science* 126 (23 August 1957): 335–43. Note also Ralph A. Morgen, "The Place of the National Science Foundation in the Basic Research Program," *Proceedings of the 6th Annual Conference of the Administration of Research* (Engineering Experiment Station, Georgia Institute of Technology, 1953), 48–54 and "NSF Program," *Science* 114 (23 November 1951): 538.

[69] On the history of NSF, see J. Merton England, *A Patron for Pure Science: The National Science Foundation's Formative Years, 1945–1957* (Washington, D.C.: National Science Foundation, 1982); Dorothy Schaffter, *The National Science Foundation* (New York: Frederick A Praeger, 1969); Milton Lomask, *A Minor Miracle: An Informal History of the National Science Foundation* (Washington, D.C.: National Science Foundation, 1976); Michael D. Reagan, *Science and the Federal Patron* (New York: Oxford University Press, 1969); Detlev W. Bronk, "The National Science Foundation: Origins, Hopes, and Aspirations," *Science* 188 (2 May 1975): 409–20; and especially Detlev W. Bronk, "Science Advice in the White House," *Science* 186 (11 October 1974): 116–22.

CHAPTER FOUR

The Office of No Return? ONR and the Issue of Relevance

THE GOLDEN AGE of academic science in America was quite brief. Between 1946 and 1950, ONR was able to support, with public funds, promising opportunities in science, largely free from the consideration of the geographic distribution of the funds and the expected public benefits of the science. Government in America had traditionally been indifferent to the interests of science, supporting only those research projects or fields where the political and economic returns were obvious.[1] During these four years, however, it seemed as if the discovery of new knowledge was coming to be valued in America for its own sake. But after 1950, although ONR and other agencies of government would continue to finance science, this support would no longer be free from the pressure for relevance.

Many mistakenly believe that the resurgence of a concern for relevancy in government support of science dates from the adoption of the so-called Mansfield amendment that was attached to the fiscal year (FY) 1970 Military Procurement Authorization Act, and which required that all defense-sponsored research have a direct relationship to a specific military function.[2] The Mansfield amendment, in fact, marks only the return of a congressional interest in the relevancy of research. The executive branch, at progressively earlier stages as one moves down the hierarchy, had previously indicated the same interest. In 1966 President Johnson in a statement specifically referring to biomedical research, but interpreted to be of general applicability, noted that society expected visible benefits from its investments in

[1] This attitude is well illustrated in the history of the Geological Survey, the largest scientific agency of the federal government in the 19th century. Thomas G. Manning, *Government in Science: The U.S. Geological Survey 1867–1964* (Lexington, Ky.: University of Kentucky Press, 1967); William Culp Darrah, *Powell in Colorado* (Princeton, N.J.: Princeton University Press, 1951). See also my "Science Policy," in F. Greenstein and N. Polsby, eds., *Handbook of Political Science* (Reading, Mass.: Addison-Wesley, 1975), 6: 79–110.

[2] Note J. Thomas Ratchford, "Congressional Views of Federal Research Support to Universities," a paper delivered at the 1972 meetings of the American Association for the Advancement of Science, December 1972. The Mansfield Amendment is historically mistitled, for it was Senator William Fulbright rather than Senator Mansfield who actually introduced the amendment for Senate debate.

basic science.[3] Over a decade earlier, the Secretary of Defense, Charles E. Wilson, announced that it was not the intention of the Department of Defense to support pure scientific research, or as he characteristically put it, "research on why potatoes turn brown when they are fried."[4] And in 1950 the Navy rediscovered the Office of Naval Research.

Politically sophisticated scientists, Warren Weaver certainly among them, knew there would be a time of reckoning with the Navy, and tried to prepare ONR for the inevitable trauma. At the spring 1947 meeting of the Naval Research Advisory Committee, Weaver, the committee's chairman and a long-time foundation executive, interrupted the routine review of the minutes of the previous session to propose a correction. The record had included a statement by another scientist which read, "ONR could not have been more generous in [its] support of research." Weaver suggested "more effective" be substituted for "more generous," phrasing which he thought was subject to misinterpretation.[5] Although the correction was adopted, such subtlety was unneeded during the late 1940s, as ONR at that time operated apart from the rest of the Navy.

In the months immediately preceding the Korean War, however, the Navy began to feel the effects of the budget cuts President Truman had imposed on the Department of Defense. A search was initiated within the Navy for nonessential activities that could be eliminated. ONR's university research program appeared to senior naval officers to be an obvious candidate for termination. In January 1950 the Chief of Naval Operations assigned Vice Adm. Oscar C. Badger, a Medal of Honor winner and an officer without previous experience in research, to review the program for possible savings.[6] Although ONR survived Admiral Badger's review, only the timely start of the Korean War prevented its demise. ONR could never again forget it was a naval organization.

[3] President Johnson's statements on research productivity and the politics that lay behind them are discussed in John Walsh, "NIH: Demand Increases for Applications of Research," *Science* 153 (8 July 1966): 149–52 and Daniel S. Greenberg, "L.B.J. at NIH: President Offers Kind Words for Basic Research," *Science* 157 (28 July 1967): 403–9.

[4] The fried potatoes quote and Secretary Wilson's attitudes toward science are reported in "Wilson Hits Generals for Opposing Aid Cuts," *New York Herald Tribune*, 9 June 1953 and "Wilson to Oppose Military Cutbacks After Korean Truce," *New York Times*, 9 June 1953, 1, 8. The Secretary was asked to clarify his views on research at his next two news conferences (Washington, D.C. on 16 June 1953, and Quantico, Virginia on 21 June 1953).

[5] Transcript, third meeting of the Naval Research Advisory Committee, Washington, D.C., 26 March 1947.

[6] Transcript, eleventh meeting of the Naval Research Advisory Committee, Washington, D.C., 30 January 1950.

THE BADGER REVIEW

There was always another ONR, an ONR deeply involved in supporting research of obvious relevance to the Navy. The agency sponsored, for example, both the Naval Research Laboratory, which had long been a leader in warfare-related innovation, and the Special Devices Center, which nominally had been established to help improve the Navy's training capabilities, but which apparently devoted at least some of its effort to developing gadgetry for the nation's intelligence services. In addition, there were from the beginning, officers and civilians within ONR's Research Group whose only interest was in promoting advances in naval weapons. Small groups of them worked to encourage research on sonars, missiles, ship propulsion, and aircraft systems and components.

The prime focus of ONR's activities during the immediate postwar years, however, was promoting advances in academic science. The Contract Research Program, which absorbed half of ONR's budget and most of its staff's attention, was almost entirely directed toward the support of basic research projects initiated by university-based investigators. It was this program that gave ONR its esteemed standing among scientists. And it was this program that was vulnerable to scrutiny by senior naval officers determined to save resources for a financially strapped fleet.

In late 1949 the Navy was desperately searching for ways to reduce expenditures. It had just lost a bitter political struggle with the Air Force and the Truman administration over the structure of U.S. forces, the famous "Revolt of the Admirals," which took place during the summer and fall of 1949. The leadership of the Navy had advocated, too vocally it turned out, the construction of a super carrier in order to gain for the Navy a share of the strategic bombing mission, the mainstay of U.S. defenses at the time. The administration decided instead to build a force of B-36 bombers for the Air Force, taking part of its cost out of the Navy's budget. Several senior officers including Admiral Denfeld, the Chief of Naval Operations, were forced to retire early because of their public protest of the budget cuts and the dashing of the Navy's strategic ambitions. Their replacements had no choice but to implement the assigned reductions which were to total $376 million.[7] ONR's Contract Research Program, budgeted at $50 million for FY 1951 (including $20 million allocated for contract longevity), was among the first of the many shore support programs to at-

---

[7] The entire incident is described in Paul Y. Hammond, "Super Carriers and B-36 Bombers: Appropriations, Strategy, and Politics," in Harold Stein, ed., *American Civil-Military Decisions* (Birmingham, Ala.: University of Alabama Press, 1963), 467–89.

tract the attention of the Navy's high command, as its continuance did not appear vital to the maintenance of the fleet's combat readiness, which had to be protected even in a period of fiscal austerity.

In preparation for a revision of the Navy's FY 1951 budget, Vice Adm. Oscar C. Badger was assigned by the new Chief of Naval Operations, Admiral Sherman, to determine the savings that might be gleaned from the program. Given Admiral Badger's reputation as a hard-bitten, combat-oriented officer, most members of the ONR staff expected the worst. They were certain that the Admiral was sent to close the office, no doubt at the urging of the materiel bureaus which viewed ONR as a budgetary rival in hard times.

Morale was so shattered when Admiral Badger arrived in late January 1950 to review the Contract program, that he continually had to reassure the staff of his intentions. He was there "to help save the Navy a little money," but not "at the expense of hurting the Navy."[8] No one could be more appreciative of the value of ONR's activities than he was, though he was worried about the importance of some of its projects. Of course, he did not know much about research, but perhaps he could offer some constructive recommendations despite this deficiency.[9] Some of his best friends were scientists, he said.[10] The Admiral then began a project-by-project review of the Contract program, that was remembered by the staff as an inquisition.[11]

The Admiral conducted the review in the form of a hearing, seated at a table with his dog loyally lying at his side as the branch chiefs attempted to justify each of their projects that the office supported. Testifying during the day, the staff worked each evening refining project rationales. An argument that once aroused the Admiral's anger was not used again. He had no patience, it was learned, for economic or political justifications. An attempt to describe the support of research as a public relations device useful in currying favor with scientists was brushed aside icily by the Admiral; the mention of the possibility that a particular project might have benefits for the civilian economy drew only the comment, "Leave Mrs. Badger out of this."[12] It was the value of basic research to the operational capabilities of the Navy that Admiral Badger sought to determine.

Soon each project had acquired a thick coat of naval rationales, too

---

[8] Eleventh meeting of the Naval Research Advisory Committee, Washington, D.C., 30 January 1950, 162.

[9] Ibid., 157.

[10] Ibid., 20.

[11] Interview documents.

[12] Eleventh meeting of the Naval Research Advisory Committee, 15, and Interview document.

thick perhaps. For example, faces surely reddened when the psychology division described a study of leadership traits as an examination of naval officers just like the Admiral so that the Navy could recruit the best officers.[13] Better received were the references by the physics division of the important role nuclear weapons were likely to play in future conflict, and the geography division's description of the casualties that ignorance of terrain features cost the amphibious landings at Salerno during the Second World War.[14]

The Admiral was a convert by the end of his two-week inquiry. The report he submitted, which ONR staff helped write, stated, "Within the limits of approved policies, directives, and availability of funds, the *research programs* in universities of the Bureaus and the Office of Naval Research *are on a sound and objective basis* and are controlled, administered and continually reviewed and adjusted with a high degree of efficiency and appreciation of the overall objectives" (italics in the original).[15] ONR appeared to have survived its most threatening bureaucratic challenge.

However, the danger had not totally passed. The General Board of the Navy, at the time the service's highest policy review body, took a less sanguine view of ONR's contract research program than did Admiral Badger. Its opinions were expressed in a report completed one month after Badger's.[16] To begin with, the Board felt that ONR was not paying sufficient attention to naval needs in its research planning. One way to achieve this attention that the Board wished considered, was to have supervisory responsibility for the office shifted from the Navy's Secretariat to the Chief of Naval Operations, thus handing ONR over to military control.[17] In addition, the Board thought that there was still room for savings in ONR's support of university research. "There is disagreement as to the justification for the amounts now spent by the Navy for basic research, which is currently about 10 per-

[13] Interview documents.

[14] Interview document.

[15] The official citation for the report is V. Adm. Oscar C. Badger sec. ltr. OCB/tn Serial: 001 of 13 February 1950 to CNO. Subj: Investigation of Navy Research and Development contracts with universities and colleges; report of (8 encls.). Despite extensive searches, the report has not been located in Navy files. It is, however, quoted at length in the progress report of the Kendall Board; see n. 20 below. See also Interview document.

[16] Report of the General Board on the Relationship of Various Budgetary Programs to Maintain a Most Effective Navy, 13 March 1950, General Board Files, Office of Naval History.

[17] Notes by N. J. Frank, Secretary to the General Board, on a meeting of the board with Capt. G. C. Wright, OP-34, Project 8-49, 3 March 1950, General Board Files, Office of Naval History.

cent of the research budget. The Board recognizes the necessity for basic research in the overall scientific development of the United States. However, expenditures for this purpose should be assigned a relatively lower priority if further curtailment of the total research and development budget is necessary."[18]

The Chief of Naval Operations also seemed dissatisfied with Admiral Badger's assessment of the naval relevance of ONR's contract research program. Less than two weeks after receiving his report, Admiral Sherman directed Rear Adm. H. S. Kendall to head another inquiry, one that would seek opportunities for a closer integration of research and development activities in the Navy.[19] Once again ONR's concern for basic research seemed about to be placed in jeopardy.

But before Admiral Kendall's study group could turn to a detailed examination of the ONR contract research program, the Korean War broke out.[20] Admirals Badger and Kendall, along with the rest of the Navy, shifted their attention to the Pacific and the fighting. Soon there was money for every type of research protected by a defense rationale. Because of the North Korean attack, ONR survived its Badger/General Board crisis, but not without a permanent concern for the relevance of its research.[21]

THE SEARCH FOR A NAVAL MISSION

The Korean War increased the pressure on ONR for relevance because it precipitated the rearmament of America, so great was the public's perception of the global threat posed by this act of Soviet-inspired expansionism. Military-related activities in the United States had a sense of urgency lacking since the end of the Second World War. Government agencies unable to adjust their operations to the emergency were certain to be restructured or bypassed. In the case of defense research, there were proposals to establish an OSRD-type organization and to centralize research programs within the services. Science was going to be mobilized with or without ONR.

The agency's response was twofold. On the one hand, it became

[18] Report of the General Board, 13 March 1950, 59.

[19] Letter from CNO to R. Adm. H. S. Kendall, 24 February 1950, CNO files, Office of Naval History.

[20] The Kendall Board, officially the U.S. Navy R&D Survey Board, did issue one progress report before disbanding, in which it quoted approvingly from the Badger Report, but in which it also noted that there should be a careful review of basic research being conducted by ONR and the Navy. U.S. Navy Research and Development Survey Board, Progress Report, 1 July 1950, CNO files, Office of Naval History.

[21] See Luther Carter, "Office of Naval Research: 20 Years Bring Changes," *Science* 153 (22 July 1966): 397–400.

much more politically sophisticated than it had been about the presentation of its basic research activities. No longer did it seek to justify the support of research in terms that would appeal largely to scientists. Instead, ONR developed rationales for this work that would link basic research directly to current and projected naval warfare needs. On the other hand, it did begin to alter the content of the research in order to become more involved than it had been in the development of naval weapons. Gradually, the ONR that had been totally devoted to the support of academic science merged with the Navy-oriented ONR. As a consequence, there was an even more intense search for an all-encompassing naval mission for the organization.

The new-found political sophistication was reflected in a number of ways. For example, ONR became a master of the so-called "two title policy." During his inquiry, Admiral Badger had gotten hold of the ONR project books that contained project descriptions prepared by scientists who were receiving funds and had been upset that these descriptions did not show a concern for naval needs.[22] From then on, two titles and descriptions were prepared; one by the scientist directing the project, and the other, much more graphically military, prepared by ONR staff.[23] Projects were "painted blue" so as to better survive internal government reviews.[24] An investigator would think he was working on "High-power broadly tunable laser action in the ultraviolet spectrum" to take an example from electrical engineering, but within the Navy the project's title was "Weaponry—lasers for increased damage effectiveness."[25] The development of scientific computational machines became the study of command and control systems, and exploration of Pacific Island cultures became the study of favorable amphibious assault sites. As one ONR division chief later confided, much creativity was expended in preparing defensible project titles and descriptions. "I hope," he said, "that St. Peter will forgive me for some of the justifications I dreamed up."[26]

The Office also began to back away from controversial fields of research, especially the social sciences, which might increase its vulnerability. Admiral Badger was not the only naval officer who felt uncomfortable with the Navy supporting psychological or foreign policy studies. No one on the staff was ready to risk the organization for the

---

[22] Fifteenth meeting of the Naval Research Advisory Committee, Washington, D.C., 25 June 1951, 24.

[23] Deborah Shapley, "Defense Research: The Names Are Changed to Protect the Innocent," *Science* 175 (22 February 1972): 866–68. See also Interview document.

[24] Interview document.

[25] Shapley, "Defense Research," 866.

[26] Interview document.

sake of maintaining its pioneering role in these and similar disciplines. Because their rationales were more readily accepted, research projects that emphasized potential hardware applications gained priority.[27]

Finally, ONR made certain that its relations with Congress were in good order. It perfected for presentation to congressional committees an annual "dog and pony" show that emphasized the naval weapons applications of research and which was well laced with attention-catching visual aids. Invariably included were ship models depicting the latest designs and film clips of attacking aircraft and missile launches.[28] Congressman Shepard of the House Appropriations Committee was one member at least who was impressed by the effort, for he was quoted as saying, "I always look forward to the Office of Naval Research presentation because you fellows always bring over such interesting things to show us."[29]

But despite its sophistication, ONR could not totally resist the pressures to redirect its research support toward more applied activities. Money became available for weapons-related projects, but not for additional basic research support. The question that kept getting asked was what ONR was doing for the Navy.

Pressure came from within the organization as well as without, from friends as well as those who wished its curtailment or elimination. In November 1950, Emmanual Piore, ONR's Chief Scientist, invited the Bird Dogs (the young reserve officers who had helped found the Office) to review its operations in light of the Korean War–induced emergency and to suggest improvements. Their report, completed during February 1951, argued for direct ONR involvement in the planning and coordination of naval research. Although ONR, in their view, had done extremely well in promoting academic research, it had failed to link that work to the needs of the Navy. The law establishing the Office, they noted, had in effect turned the Chief of Naval Research into the Navy's "Vice President in Charge of Research." They thought it was time that this responsibility was exercised, and proposed that the most effective way to do this would be to assign the Chief of Naval Research additional duty in the Office of the Chief of Naval Operations, thus binding ONR to the fleet.[30]

Rear Adm. Thomas Solberg, then Chief of Naval Research, was to-

[27] Interview documents.

[28] Interview documents.

[29] Admiral Furth quoting Congressman Shepard, twenty-second meeting of the Naval Research Advisory Committee, Washington, D.C., 18 March 1954, 8.

[30] J. T. Burwell, R. A. Krause, B. S. Old, and J. H. Wakelin, "Recommendations on the Office of Naval Research," 15 February 1951 (unpublished report to ONR).

tally opposed to the report's main recommendation.[31] He knew that closer ties to the Chief of Naval Operations would undermine ONR's independence and would threaten its basic research program. ONR had flourished because it reported to the Secretary and not to the Chief of Naval Operations. Moreover, he doubted that ONR could effectively coordinate naval research when it itself was one of the several agencies within the Navy supporting research. The NSF could not accomplish this task for the government as a whole, and neither could ONR for the Navy. But in successfully resisting the report's conclusion that reorganization was needed, Admiral Solberg had to claim that ONR was already beginning to fulfill its role as the manager of the Navy's research activities. He argued that ONR was planning research for the entire Navy and coordinating its program with those of the Navy materiel bureaus.[32] The very mission that Admiral Solberg sought to avoid became the first ONR asserted as uniquely its own.

Unintentionally perhaps, Congress soon gave a boost to ONR's ambition. Until the early 1950s, there was no separate research and development (R&D) appropriation for the Navy. Instead, R&D allocations were mixed within the procurement appropriations of the materiel bureaus, permitting them to ignore congressional intent by shifting monies without legal restraint from R&D activities to weapon procurement activities and vice versa. Only ONR's budget was separately identified as research. In order to bind the bureaus to congressional will, Congress mandated that beginning with the FY 1954 budget, an appropriation category labeled "Research and Development, Navy" be established that would consolidate all R&D appropriations for the Navy into a single account, nontransferable to other activities without congressional approval. ONR was designated as the account coordinating agency within the Navy. Although this responsibility was primarily ministerial in nature, it did give ONR the opportunity to monitor the R&D programs of every organizational element within the service.[33]

Following closely on this congressional action was a decision by the

---

[31] Interview documents.

[32] Memorandum to Adm. L. D. McCormick from Rear Adm. T. A. Solberg. Subject: report by Drs. Burwell, Krause, Old, and Wakelin, dated 15 February 1951, report enclosed, 27 March 1951. Most revealing are Admiral Solberg's marginal comments on the Bird Dogs' report. Bruce Old continued his work on R&D management and especially on the contribution of basic research throughout his career. See "What Good is Basic Research? Bruce Old Tries a New Method and Gets a Big Answer," *Technology Review* 91 (August/September 1988): A10–A11.

[33] Booz Allen & Hamilton, Inc., *Review of Navy R&D Management 1946–1973*, 1 June 1976, Summary Volume, 38–39; transcript of the twenty-fourth meeting of the Naval Research Advisory Committee, Washington, D.C., 1 December 1954.

Secretary of the Navy to expand the scope of ONR's coordinating responsibilities. In April 1954 the Under Secretary of the Navy, Thomas Gates, reported to Secretary Robert Anderson on his study of the organization of the Navy. Gates's study had been undertaken to determine what adjustments, if any, were necessary in the Navy's organization as a consequence of the restructuring of the Department of Defense that occurred in 1953, and of the growing congressional interest in the management of defense projects.[34] One area of sensitivity was the organization of R&D activities. In response to congressional pressure, both the Army and the Air Force had already established an Assistant Secretary for R&D.[35] Gates argued against such a post for the Navy, preferring to leave supervisory responsibility for the Navy's nearly half billion dollars in R&D activities with its Assistant Secretary for Air. Instead, he proposed that the Chief of Naval Research, who already reported to the Assistant Secretary, be assigned responsibility for coordinating all of the Navy's R&D programs, a recommendation the Secretary accepted.[36]

With the Chief of Naval Research's new responsibility, ONR had obtained a mission that brought it directly to the center stage of naval politics. The coordination of development programs required the choice among technologies, the selection of weapon systems to be readied for procurement. Because the Navy is built around competing technologies (aircraft, surface ships, and submarines to name the three most important), this choice necessarily influences the distribution of power within the Navy and is subject to intense bureaucratic conflict. The coordination of research was a difficult enough task for ONR, given its relative weak political base, academic scientists, and a few converted naval officers. Now it was being asked to adjudicate the Navy's future structure with nothing additional in its favor other than a directive from the Secretary of the Navy.

A coordinating office was established within ONR to conduct analyses of developmental options, apparently in the hope that there were rational choices to be made among the options. Periodic reviews of the operations of the office noted that it was undermanned for the task, but no matter what staffing levels could be obtained, the mission itself was doomed.[37] ONR and the Chief of Naval Research simply had in-

---

[34] Robert J. Mindak, *Management Studies and Their Effect on Navy R&D*, Office of Naval Research, 1 November 1974, 4–5.

[35] Ibid., 6.

[36] *Report of the Committee on Organization of the Department of the Navy*, Thomas S. Gates, Chairman, 16 April 1954.

[37] See, for example, *Research and Development in the Government*, a report to Congress by the Commission on the Organization of the Executive Branch of Government,

sufficient standing within the Navy to impose order on the vast number of development projects and the interests they represented.[38]

The Office's senior staff recognized that the organization could not succeed in its coordinating role.[39] Led by Emmanual Piore, they sought to turn the situation to ONR's advantage by having the agency take the initiative in advancing high-risk technologies with large potential application to naval weapons such as sensors and missiles. This thrust into exploratory development was likened by one participant observer with a sense of history to Sweden's attempt to become a world power under Gustaf Adolphus.[40] And similar to Sweden's European ventures, it was frustrated by coalitions of threatened powers. In ONR's case, the materiel bureaus claimed these technologies as exclusively within their own jurisdictions no matter how slowly or ineffectually they were exploiting them. At that time not a single, technological foothold was ceded to ONR.

In 1959 another departmental reorganization study group, the Franke Committee, recommended the removal of development coordination authority from the Chief of Naval Research and its assignment to a Deputy Chief of Naval Operations, a three star admiral in direct line to the Chief of Naval Operations.[41] The Secretary, now Thomas Gates, approved the transfer of jurisdiction and some personnel to the Chief of Naval Operations Staff (OPNAV), leaving ONR responsible only for the administration of the department's R&D appropriation account and the support of most, but not all, of the Navy's basic research and the level of applied research that would complement the work of the bureaus. Although the new arrangements did not solve the development coordination problem within the Navy (there were to be nine Deputy Chiefs of Naval Operations for Development appointed in the next thirteen years), it did remove ONR from the continuing conflict over the Navy's technological priorities.[42] ONR had failed with the one naval mission that the Navy had assigned the

---

Herbert Hoover, Chairman, May 1955. Task Force Subcommittee Recommendation No. 11. See also E. R. Piore, "Office of Naval Research, The First Ten Years: A Summary," ONR, circa 1955.

[38] Interview documents.

[39] E. R. Piore, "Office of Naval Research, The First Ten Years: A Summary."

[40] Interview document.

[41] *Report of the Committee on Organization of the Department of the Navy*, William B. Franke, Chairman, 1959, recommendation C-2. Mindak, *Management Studies*, 7. Ironically, the Franke Board also recommended the appointment of an Assistant Secretary for Research and Development, a post that Secretary Gates had resisted in 1954, but one that he now quietly accepted.

[42] Interview document.

organization. Given the nature of that mission, it could not have done otherwise.

ALWAYS ANOTHER CRITIC

Except for brief reprieves invariably associated with national crises, ONR's support of basic research has been under nearly constant political attack since the early 1950s. Although the source of the criticism has kept changing, its point has not. Again and again, ONR has been forced to defend the belief that such support is in the interest of the Navy and the nation. Each of the major confrontations on the issue has had the effect of reducing the program, curtailing either its scope or relative scale, if not always the base allocation.

Because of his unusually direct manner of expression, the most colorful of the many antagonists surely was Charles E. Wilson, President Eisenhower's first Secretary of Defense who held that post until the Sputnik crisis in 1957. The Eisenhower administration had taken office with the pledge to control government expenditures, which at the time necessarily meant it had to seek reductions in defense spending. Among the prime targets for budget cuts was military research and development.[43]

A former Chairman of the Board at General Motors, Secretary Wilson, had definite ideas about the value of basic research which he was quite willing to express publicly. In one of his first congressional appearances, Wilson recalled that Charles Kettering, a colleague at General Motors and one of the nation's greatest engineers, defined pure science as science that "if successful, it could not be of any possible use to the people who put up the money for it—that made it pure." The Secretary then went on to make his famous fried potatoes statement to underline his determination not to waste defense dollars on basic research.[44] Less than two weeks later, Wilson announced additional cuts of 25 percent in the Defense Department's R&D expenditure plans, already substantially reduced from the amounts proposed by the Truman administration before its term expired.[45]

Of course, budget reductions of the levels Secretary Wilson sought are difficult to achieve. Congressmen and the armed services rushed to protect favored projects and the civil servants and military personnel they supported. Thus, Navy R&D as a whole and ONR as an orga-

[43] "Wilson Suspends Build-up Targets," *New York Times*, 13 May 1953, 1.

[44] "Wilson to Oppose Military Cutbacks After Korea Truce," *New York Times*, 9 June 1953, 1.

[45] "Wilson Orders New Slash of 25% in Research Fund," *New York Times*, 18 June 1953, 27.

nization sustained reductions of only 16 percent in their FY 1954 allocations. Navy sponsorship of university research, mostly provided by ONR, declined by the full 25 percent, however. The Contract Program, which had been as high as $50 million two years before, fell to less than $30 million, a level at which it was to languish through Wilson's entire tenure as secretary.[46]

Wilson had no objection, as he repeatedly stated, to other government agencies such as NSF or NIH supporting undirected basic research. It was only the defense budget that he insisted be lean.[47] Yet, Congress was not then ready to make up the difference in university research support eliminated in defense reductions by providing compensating increases in the budget of the NSF, the most appropriate alternative sponsor for the research affected. On the contrary, Wilson's rhetoric merely stimulated congressmen to provide their own evaluations of the benefits of university-based research. Senator John McClellan, for example, apparently thought an Air Force–sponsored study at Harvard on socialism in the Soviet Union did not merit anyone's support. "What do you get [for this]," he asked a general, "just a lot of professor theories and all that stuff? Is that what you get out of it? To me that is simply throwing money away, nothing else."[48] Coupled with the witch hunt for Communists stimulated by Senator Joseph McCarthy, the budget reductions had a sobering effect on the universities and their friends in government.

To be sure, loyalists at ONR tried to resist the political tide, but without success. A letter drafted for NRAC to send to Secretary Wilson claimed (incorrectly) that the committee had studied the impact of the budget cuts on ONR and found that they would adversely affect its ability to provide the necessary technical and engineering knowledge needed to maintain future naval power.[49] Some of the longevity funds acquired in the late 1940s were expended in order to reduce the severity of the reduction in the Contract Program.[50] At the same time, a

[46] National Science Foundation, *Federal Funds for R&D, 1953–1980* (Washington, D.C.: Government Printing Office, 1981); R&D Comptroller, Office of Naval Research; transcript of the thirty-fourth meeting of the Naval Research Advisory Committee, Washington, D.C., 11 July 1957.

[47] See, for example, transcript of Secretary Wilson's news conference, Quantico, Va., 21 June 1956 and "Soviet Airpower Held Overrated," *New York Times*, 27 February 1957, 13 where the Secretary expressed his desire that funds for "abstract work" be spent by agencies other than the Department of Defense.

[48] Quoted in Clifford Grobstein, "Research and Development: Prospects, 1954," *Bulletin of the Atomic Scientists* 9, no. 8 (October 1953): 302.

[49] Transcript of the twentieth meeting of the Naval Research Advisory Committee, Washington, D.C., 16 June 1953.

[50] Interviews.

new rationale for basic research was invented; it was now supported to provide a "listening post" in fields of science that could later become vital to the Navy.[51] All of this notwithstanding, however, it was still necessary to modify university-supported research toward more obviously naval interests, as one senior ONR official described, "Just to reduce the amount of badgering and to be in a strong position of argument."[52] Basic research would be supported, although now with a more naval orientation than had been the practice.

But the badgering did not diminish. President Eisenhower remained concerned with the level of defense expenditures, swollen in the mid-1950s by missile and space programs including the Vanguard satellite project managed by the NRL. Significant reductions in personnel and research were imposed in 1957 and additional cuts were planned for the following year.[53] Before the impact of these efforts at expenditure control were fully felt, however, the Soviet Union launched Sputnik, their unmanned space satellite. The President and defense officials were then forced to explain to a frightened public and a disbelieving Congress that American military defenses had not been jeopardized by the Soviets' stunning technological accomplishment. Billions of additional dollars flowed into the Department of Defense just to make certain that was the case. Anxieties only increased when the ONR-sponsored Vanguard satellite failed spectacularly in its initial test.

Scientists were unanimous in attributing the blame for the Soviet lead in space, and by implication strategic missiles, to inadequate support for basic research and graduate education. There was but one direction for policy to take. Within weeks of Sputnik I, NRAC enthusiastically endorsed an ONR proposal for a 40 percent increase in the agency's allocations to academic science.[54] A year later, the President's reconstituted science advisory committee in its first public report argued that research and educational investments minimize the

[51] Transcript of the thirty-fourth meeting of the Naval Research Advisory Committee, Washington, D.C., 11 July 1957.

[52] Transcript of the twenty-first meeting of the Naval Research Advisory Committee, Washington, D.C., 17 November 1953, 71.

[53] Transcript of the thirty-fourth meeting of the Naval Research Advisory Committee, Washington, D.C., 11 July 1957, 78 and "Navy Believes Cost of R&D Can Be Cut," *Aviation Week*, 3 June 1957, 131–37.

[54] Transcript of the thirty-fifth meeting of the Naval Research Advisory Committee, Washington, D.C., 29 October 1957. Despite one member's admonition that the number should not be pulled "out of the air," it is obvious from the discussion that none of the members present had a clear notion of the basis for the percentage increase requested. It was thought a number, as another member suggested, that people swallow easily.

need for crash programs to redress technological imbalances. "It is difficult to imagine more fruitful and prudent ways to spend the taxpayer's money than on basic research," the report concluded.[55]

For a while at least, the budget restraints on basic research were in fact relaxed. ONR's Contract Program, its main mechanism for providing university support, leaped first by $20 million and then again by more than $30 million. What was once a $30 million program in FY 1957 became an $82 million effort by FY 1960.[56]

Hoping to stimulate further the new-found official willingness to support science, NRAC had ONR commission A. D. Little, Inc. to conduct a study of the potential contribution of basic research to the Navy's R&D effort. The study (led by Bruce Old, the former Bird Dog and now a vice president of A. D. Little) not surprisingly saw severe consequences for the development of future naval weapons if basic research were ever to be neglected again. The major industrial corporations, it was noted, were already substantially increasing their investments in fundamental studies, some supposedly allocating 15 percent or more of their R&D budgets to basic research. The study concluded that the appropriate share for basic research in the Navy ought to be 15 to 20 percent, a level of effort that would necessitate a doubling of the already enlarged post-Sputnik allocation for basic research in the Navy.[57] Although appreciative of the advice, the Navy's senior officials were appropriately cautious in committing themselves to earmarking any portion of their R&D effort permanently to basic research. Yes, basic research deserved support, but perhaps not a blank check, was their position. As Secretary of the Navy Franke stated at a news conference when the study was released to the public, there were dangers in "fixing any part of the budget at an arbitrary percentage."[58] Nevertheless, it seemed as if the ghost of Charles E.

[55] *Strengthening American Science*, a report of the President's Science Advisory Committee, 27 December 1958. Emmanual Piore, then recently of ONR, served as chairman of the reporting panel.

[56] The Eisenhower administration was somewhat reluctant to push ahead with increased basic research support. As the ONR comptroller reported, "The newspapers and Congress gave us a lot of money but actually at the working level we had little extra. It was more than six months after Sputnik before the first supplemental appropriations reached ONR and then for only $14.2 million. Additional money came in later, boosting the FY 58 allocations by at least $20 million. Further increases in FY 59 and 60 provided the $50 million total." Transcript of the thirty-seventh meeting of the Naval Research Advisory Committee, Norfolk, Va., 21 April 1958, and ONR budget data.

[57] "Beef Up Navy Research," *Chemical and Engineering News* 37 (26 October 1959): 28.

[58] "Scientists Urge More Navy Funds," *New York Times*, 22 October 1959, 11.

Wilson had been expunged from the Pentagon. There was money again for basic research.

Sputnik brought not only money, but also reorganization for defense and science agencies, which in turn led to the ensuing confrontation over basic research. The post of Director of Defense Research and Engineering was established to manage the department's expanded R&D activities. In addition, a new organization, the Advanced Research Projects Agency (ARPA), was created within the Office of the Secretary of Defense to provide support for research of broad interest to the services or for which no service was willing to provide funding. Immediately, ONR partisans saw the twin threats of centralization and consolidation in these changes.[59] ARPA might dictate research directions and eventually absorb the service units supporting basic and applied research. Despite some temptations to exercise the newly acquired power, the Eisenhower administration continued to tread lightly on research.[60] However, its term in office was soon coming to an end.

The Kennedy administration, given its multitude of ties to the academic community, appeared predisposed to the support of basic research. Its rhetoric was certainly unfailingly favorable to such support.[61] But under the leadership of Secretary Robert McNamara, the management of the Department of Defense became ever more hostile toward the science ONR wanted to pursue. There were new budget classifications imposed that emphasized linkages of research to military missions. Decision making within the Department of Defense became more centralized, as did the internal command structure of the services. And, worst of all from the ONR perspective, cost effectiveness analysis was applied to research, which meant that there was a continuing demand on ONR to attempt to quantify the expected benefits of its research investments. Paperwork increased while the research budget stagnated.[62]

With the stagnation in its budget, a malaise overcame the organization. Opportunities appeared to lie elsewhere and many of the more ambitious civilians moved on to other jobs. The senior staff members

[59] Transcripts of the thirty-sixth and thirty-seventh meetings of the Naval Research Advisory Committee, Washington, D.C., 22 January 1958 and 21 April 1958. See also Interview document.

[60] "Basic Research in the Defense Department: The Department's View," *Science* 132 (8 July 1960): 75–77; transcript of the thirty-eighth meeting of the Naval Research Advisory Committee, Washington, D.C., 29 July 1958, 18.

[61] Note, for example, the Bell Report (Bureau of Budget, Report to the President on Government Contracting for Research and Development, 1962).

[62] Interview document.

who remained attempted to revitalize ONR by seeking a role in the Navy's ballistic missile program through studies of long range, seabased deterrent forces.[63] This again thrust ONR into the central political arena of the Navy from which it was soon ejected once more. The important mission of determining the Navy's strategic doctrines and hardware was claimed by other, more powerful groups.[64]

The malaise was officially recognized in a discussion paper entitled "Changing Times," prepared in September 1967 by the then newly appointed Chief of Naval Research, Rear Adm. Thomas B. Owen.[65] "Changing Times" catalogued the numerous environmental shifts affecting ONR, including the altered structure of the Department of Defense, President Johnson's desire for greater application of research results, and the war in Vietnam, and was in essence another plea for the discovery of a viable mission for the organization within the Navy.

Unmentioned in "Changing Times" was growing opposition on university campuses to the war and the consequences that that might hold for military-sponsored research. It was the side effects of the war in Vietnam, rather than the war itself, that would have the greatest impact on ONR during the next several years. Because of the disruptions on the campuses, especially those directed toward the Reserve Officer Training Corps programs, the military became increasingly reluctant to support university-based research. There was a growing opposition on the campuses to all aspects of involvement with the military including the conduct of Defense Department–sponsored research. The partnership between the universities and the military, which ONR had been so active in forming, was to be severely strained.

As the opposition to the war and the military increased, some senators and representatives began to champion the same causes within Congress. Among the vocal critics of military involvement in American society was Senator William Fulbright, Chairman of the Foreign Relations Committee. In 1969 he proposed an amendment to the FY

[63] Transcript of the fifty-first meeting of the Naval Research Advisory Committee, Washington, D.C., 13 November 1961 Interview document.

[64] Interview document. ONR was grateful for any recognition it received from the Special Projects Office, the Navy organization that developed the Polaris. Captain Wootton, ONR's planning director, congratulated the staff on the receipt of $4.6 million from the Special Projects Office in 1962, stating that it demonstrated the Office's confidence in ONR's program. ONR: 402A: ADL: adw., 13 September 1962. Minutes of Program Council Meeting Nos. 32–62, 27 July 1962. The Special Projects Office, in fact, had no interest in ONR's work and was just distributing goodwill with the allocation. When ONR began to ask for more than goodwill, it was told politely to go away. Interview document.

[65] Memorandum to Distribution List, from Chief of Naval Research. Subject: Looking Ahead, 11 September 1967, Enclosure "Changing Times."

1970 Military Procurement Authorization Act, Section 203, which stated:

> None of the funds authorized to be appropriated by this Act may be used to carry out any research project or study unless such project or study has a direct or apparent relationship to a specific military function or operation.[66]

Section 203 was narrowly adopted without debate. Subsequently, Senator Mike Mansfield, the Majority Leader, praised the action, mentioning that he had a similar view of the research appropriate for DOD to sponsor. Fulbright, correctly anticipating a difficult reelection campaign, thereupon restricted his comment on Section 203 and similar measures. Because Mansfield continued to defend the amendment, it became publicly identified as the "Mansfield amendment."[67]

Mansfield claimed the intent of Section 203 was to reduce the research community's dependence on the Department of Defense. It was time, he believed, to revise national science policy, closing down "the backdoor NSF" at the Pentagon and transferring its resources to the agency Congress had established solely for the purpose of supporting fundamental investigations.[68] In order to accomplish this, the senator insisted that the Department of Defense conduct an item-by-item review of its entire repertoire of basic research projects, a total of fifteen thousand ongoing studies involving $1.3 billion in expenditures, certifying each as either in conformance with Section 203 or not.[69]

Relevance, of course, is in the eye of the beholder. Robert Frosch, who was the Assistant Secretary of the Navy for R&D at the time, recalls that his initial review found none of the Navy's more than twenty-five hundred research projects to be vulnerable. That response, he was informed, was not acceptable. After a second review, he offered a half dozen projects for a sacrifice which was implemented.[70]

---

[66] Section 203 of Public Law 91–121, "Authorization of Appropriations for Fiscal Year 1970 for Military Procurement, Research and Development, and for the Construction of Missile Test Facilities at Kwajalein Missile Test Range, and Reserve Component Strength."

[67] Rodney W. Nichols, "Mission Oriented R&D," *Science* 172 (2 April 1971): 29.

[68] "Rechanneling the Public Resources for Basic Science Through the Civilian Agencies: A New Goal for National Science Policy," Statement of Senator Mike Mansfield to the House Committee on Science, Research, and Development, 11 August 1970.

[69] Nichols, "Mission Oriented R&D," 29; "Mansfield Says Pentagon Ignores Research Curb," *New York Times*, 26 November 1969, 1; "Pentagon Agrees to Curb Research," *New York Times*, 7 December 1969, 1.

[70] Robert A. Frosch, "Relevance, Irrelevance and General Confusion: Problems in Sci-

Officially, the impact of Section 203 was reported as minor.[71] The Department of Defense terminated approximately four hundred projects involving less than $10 million worth of research. More significant cancellations were said to have occurred because of a $64 million curtailment in defense R&D imposed by Congress the same year.[72] The wording of Section 203 was diluted the following year in the FY 1971 appropriations act, giving the Secretary of Defense such wide discretion in determining what science was relevant to defense activities as to render it meaningless.[73] The amendment lingers routinely in appropriation bills with decreased importance. Secretary of Defense Harold Brown in May 1979 approved a reinterpretation of the amendment so that the test of relevance was potential benefit to defense rather than immediate benefit.[74] By then Mansfield was part of the administration serving as U.S. ambassador to Japan and Fulbright had long been forcibly retired from the Senate.

At the operating level, however, Section 203 did have an impact. Defense research agencies were reminded once again of the political risks involved in supporting fundamental work. Ever more elaborate reviews were required to gain approval for such projects.[75] If few projects were actually terminated, it was testimony to the fact that defense research agencies, ONR included, had by then already cleansed themselves of truly basic studies which were without any military rationale. They were nearly totally committed to the support of applied research or would soon become so. The other mission-oriented agencies of government, the National Aeronautics and Space Administration and the AEC in particular, were thought to have thinned their basic research portfolios in anticipation of being subject to the same criticism that the Mansfield amendment implied, probably cancelling more projects than did the Defense Department. Although some compensating budgetary increases were given to the NSF, not all of the terminated projects found alternative support.[76]

---

ence Policy." Fifteenth J. Seward Johnson Lecture in Marine Policy, Woods Hole Oceanographic Institution, 3 January 1983, 5–6.

[71] See the remarks of Senator John McIntyre, *Congressional Record*, 28 August 1970, S14552–55.

[72] Nichols, "Mission Oriented R&D," 30.

[73] J. Thomas Ratchford, "Congressional Views of Federal Research Support to Universities," a paper presented at the 1972 meetings of the American Association for the Advancement of Science; "Mansfield Amendment Cut Down," *Nature* 228 (10 October 1970): 107.

[74] "DOD Basic Research: An Uphill Climb," *Physics Today*, January 1980, 124.

[75] Interview documents.

[76] Nichols, "Mission Oriented R&D," 30; Emilio Q. Daddario, "Need for a National Policy," *Physics Today*, October 1969, 4–9. The NSF claimed that other agencies would

More important, university scientists and engineers receiving Defense Department research support became the targets of political attack by antiwar and New Left groups. If their research support continued, it was clear, given the Section 203 requirements, then their work served a direct "military function or operation," sufficient grounds to some for condemnation.[77] Several universities established official panels, often involving students and faculty as well as administrators as members, to inquire into the appropriateness of such work at their institutions, notwithstanding the fact that nearly all defense-sponsored basic research was unclassified.[78] It was soon discovered that most of the research projects carried two titles, the military one (often not previously known to the researcher) having an unfailing tendency to link the work to some destructive military purpose such as downing aircraft or sinking ships. Many a scientist or engineer thus embarrassed, had to proclaim publicly that his or her intent would never be so malevolent.[79] Although the number of applications for defense support did not decline, the politically sensitive researcher looked elsewhere for support. Years later, the effect of the Mansfield amendment continues, especially at the elite universities where, if nothing more, political sensitivities were heightened by the experience of the 1960s and are now restimulated by President Reagan's Strategic Defense Initiative. ONR remains a source of support for university-based scien-

drop $100 million in fundamental research by the end of FY 72 as a result of Mansfield-type fears. *National Science Foundation Authorization Act of 1972*, Report of the Committee on Labor and Public Works, U.S. Senate, 92d Congress, 1st Session, no. 92–232, 22 June 1971.

[77] See, for example, Stanton A. Glantz and Norm V. Albers, "Department of Defense R&D in the University," *Science* 186 (22 November 1974): 706–11.

[78] Remarks of Dr. John S. Foster, Director of Defense Research and Engineering, before the American Nuclear Society, Seattle, Washington, 18 June 1969, 5.

[79] Senator Mansfield reprinted in the *Congressional Record* of 20 March 1970 a four-part series, originally published in *Providence Evening Bulletin*, describing the anguish of scientists and engineers over the newly found titles for their DOD research projects. See Warren H. Donnelly, "Highlight of Congressional Action on Limiting Defense Funded Research to that which has a 'Direct or Apparent Relationship to a Specific Military Function or Operation,'" Legislative Reference Service, Library of Congress, 25 March 1970, 19–29. See also James Case, "Low Profile," University of California/ Santa Barbara *Daily NEXUS* (9 November 1972) reprinted as "The University and Defense Research" in *Naval Research Reviews*, February 1973, 29–32 and Stanton A. Glantz, "How the Department of Defense Shaped Academic Research and Graduate Education," in Martin L. Perl, ed., *Physics, Careers, Employment and Education* (New York: American Institute of Physics, 1978), 109–22. There were actually three titles: one devised by the scientist; another prepared by ONR for the Navy's computerized information system (limited to twenty-six characters); and a third, also prepared by ONR, that emphasized the military relevance. Interview document.

tists and engineers, but only for those prepared to acknowledge a willingness to help it down aircraft or sink ships.[80]

DIFFERING PERSPECTIVES

In the Navy, a bureaucracy which generates intense institutional loyalty, everyone in a position of responsibility wants to believe that he or she is serving the institution's interest well, but then naturally worries whether or not others are so dedicated. As a bureaucracy, the Navy divides responsibilities between those in the organization who are to be concerned about the institution's short-range needs (in this case the ability to do combat at a moment's notice), and those who are to be concerned about its long-term viability (the ability of the Navy to survive some distant war). The continuing debate over the relevance of investments in basic research for the Navy revolves around the clash between these perspectives, impassioned by the intense institutional loyalty.

Naval officers tend to be advocates for the Navy's short-term needs even if they hold positions where the opposite perspective is more appropriate. They are, after all, the organizational members who will do the fighting if combat does occur. Moreover, they are subject to a personnel system that encourages them to press for immediate results. They are given assignments of short duration—two, three, or four years—and their promotional opportunities depend in large part on fitness reports that describe their accomplishments during these brief tours of duty. Because of their constantly changing assignments, they only rarely develop the deep expertise required for scientific standing.

The Navy's civilian scientists tend to have the opposite bias. They are socialized in their graduate training to accept as necessary the delayed gratifications inherent in the scientific process. They are taught, and most ardently believe, that all scientific knowledge is useful even though the most important applications of this knowledge may not be revealed for years. They know that significant scientific results are often painstakingly acquired. And they all arrive with, and are permitted to develop further, expert knowledge in subdisciplines whose utility to the Navy they learn becomes vital and is appreciated at only

[80] John Walsh, "Pentagon Plans Boost for Basic Research," *Science* 205 (August 1979): 566–68; Colin Norman, "Pentagon Seeks to Build Bridges to Academe," *Science* 228 (19 April 1985): 303; Paul Mann, "Defense Department Skeptical as House Panel Urges More University Research," *Aviation Week and Space Technology*, 6 May 1985, 101–2; David Dickson, *The New Politics of Science* (New York: Pantheon, 1984), 122–23.

unpredictable intervals. It is not surprising that these perspectives clash.

Most naval officers who know of ONR's existence, and not all do because of its organizational isolation and relatively small scale, believe it does not contribute appreciably to naval strength. When ONR personnel are not present, they are said to refer to the Office as the "Office of No Return," presumably reflecting both their assessments of the value of its research investments and the effect an assignment there has on an officer's career.[81]

The hostility naval officers have toward the Office is at times directly expressed. A Navy captain who was given a terminal assignment as the commanding officer of one of ONR's field offices paid a courtesy visit to the commandant of the local naval district, a rear admiral. "What is your budget?" the admiral asked. The captain replied that he had just arrived and did not yet know. "Whatever it is, it's too much," the admiral then said.[82]

Even those naval officers who volunteer for ONR duty do not always think its activities are vital to the Navy. For example, one officer who sought an ONR field assignment in order to be near his ill parents found the contract program to be a subsidy for scientists, "a giveaway program just like those for farmers." He did not enjoy handing out the money. "Scientists," he said, "are just like cats; they purr at meal times and will rub up to anyone" who might have a treat.[83] Much better he thought were the Army and Air Force research programs, which he believed were more weapons-oriented than ONR's.[84]

The civilian scientists are equally vehement in expressing their position. "The Blue Suiters," said one, "are going to ruin the Navy," because of their failure to appreciate the eventual value of scientific advances.[85] Another thought that several chiefs of naval research seeking further promotions had tried to use ONR as their private foundation to disburse money to a favorite naval charity, always one of the Navy's development agencies such as the air systems command or ship systems command.[86] A third believed that those naval officers who did not want to invest in basic research, thereby ensuring the Navy's future, "ought to be shot as traitors."[87]

Occasionally, one of the armed services or an office in the Depart-

---

[81] Interview document.
[82] Interview document.
[83] Interview document.
[84] Interview document.
[85] Interview document.
[86] Interview document.
[87] Interview document.

ment of Defense attempts to resolve with a quantitative study the debate over the appropriate division between the support for development projects (work directed specifically toward the advancement of weapons technologies), and the support for basic research (work of a more theoretical and long-term nature, not always clearly linked to military agency missions). Such was the case in the mid-1960s when the Director of Defense Research and Engineering in the Department of Defense initiated the now famous Project Hindsight study.[88] The House Defense Appropriation Subcommittee had been critical of defense research management, in particular questioning the value of substantial investments in scientific research. Intended in part as a response to that criticism, the Hindsight study was to identify the source of weapon system improvements by describing the purpose and institution location of work done up to twenty years earlier that was embodied in a sample of currently deployed weapon systems. The expectation was that the wisdom of past research investments would be demonstrated. The study did record hundreds of research events as important to the success of the weapons it examined, but hardly any were described by the investigators as being the product of basic research. What they found to be vital were highly focused efforts at the advanced engineering and early systems development stages of the R&D spectrum.

The Hindsight conclusion that basic research activities in the post–Second World War years contributed little to operational weapon systems, having as it did the implication that basic research should not be supported, led to the study being challenged by scientists on both methodological and policy grounds. It was argued that Hindsight looked at too short a historical record, ignoring in the process the contributions of basic research that underlies current development activities. The project's sample was criticized as being unrepresentative of significant advances in technology. And there was concern that because of these faults Hindsight would cause a decline in the support

[88] Memorandum from the Director of Defense Research and Engineering to the Assistant Secretaries (R&D) of the Army, Navy, and Air Force. Subj.: Project Hindsight—Expanded Studies of Research and Technology Which Have Been Utilized in Weapon Systems, 6 July 1965. See also C. W. Sherwin and R. S. Isenson, "Project Hindsight," *Science* 156 (1967): 1590–92; R. S. Isenson, "Project Hindsight: An Empirical Study of the Sources of Ideas Utilized in Operational Weapon Systems," in W. H. Gruber and D. G. Marquis, eds., *Factors in the Transfer of Technology* (Cambridge, Mass.: M.I.T. Press, 1969) 155–76; Daniel S. Greenberg, "'Hindsight': DOD Study Examines Return on Investment in Research," *Science* 154 (18 November 1964): 872–73. The actual study is: C. W. Sherwin, et al., *First Interim Report on Project Hindsight (Summary)*, Office of the Director of Defense Research and Engineering, Washington, D.C., 30 June 1966.

for basic research, and thus a retarding of future advances in technology.[89] A quickly organized counter study, Technology in Retrospect and Critical Events in Science (TRACES), sponsored by NSF, found that important and commonly used technologies (the video tape recorder, magnetic ferrites, the electron microscope, matrix isolation, and the birth control pill, but no weapon systems, were all included in the TRACES sample) had their root in scientific work of the most basic and undirected type done—in some instances, centuries earlier. The not surprising conclusion was that such research was indeed useful and worthy of support.[90]

Of course, these studies cannot resolve the dispute, reinforcing as they do the prejudices of the contending sides. The burden of adjudication lies with politicians, the senior executive branch officials, and congressmen. And given that elections occur more frequently than wars, they are even more inclined to be concerned about the short term than is the military.

Scientists who recognize the political realities respond by claiming that their work has relevance for current national problems, whatever they might be. All sorts of absurdities result. Biologists are reborn as environmentalists and physicists as energy experts, as the need arises. Surely though, the extreme was reached in the early 1970s when polar research scientists argued that their Antarctic field station experience, isolated and deprived as it always was, held lessons for the alleviation of the problems of urban ghettos.[91] Good science still gets done, but often in disguise.

In the words of John Maddox, an acute English observer, American science is involved in an endless search for objectives.[92] For a brief few years in the late 1940s, ONR was able to ignore purpose and politics as it spent a small part of the Navy's Second World War surplus

[89] Note, for example, Harvey Brooks, *The Government of Science* (Cambridge, Mass.: M.I.T. Press, 1968) 122, 291, 322; K. Krielkamp, "Hindsight and the Real World of Science Policy," *Science Studies* 1 (1971): 43–66. See also Jeffrey K. Stine, *A History of Science Policy in the United States 1940–1985*, a report prepared for the Task Force on Science Policy, Committee on Science and Technology, House of Representatives, 99th Congress, 2d Session, September 1986, 59–61.

[90] Illinois Institute of Technology, *Technology in Retrospect* and *Critical Events in Science (TRACES)* (Washington, D.C.: National Science Foundation, 1968); John Walsh, "TRACES: Basic Research Links to Technology Appraised," *Science* 163 (24 January 1969): 374–76.

[91] Panel on Biological and Medical Sciences of the Committee on Polar Research, *Review and Recommendations* (Washington, D.C.: National Academy of Sciences-National Research Council, 1971), 9.

[92] John Maddox, "American Science: Endless Search for Objectives," *Daedalus* 101, no. 4 (fall 1972): 129–40. See also W. Henry Lambright, *Governing Science and Technology* (New York: Oxford University Press, 1976) 182–209.

to revitalize American basic research. Gradually, however, the traditional American preference for pragmatic research began to reassert itself.[93] Diverting circumstance and bureaucratic subterfuge saved ONR's basic research program for a time, but substantial organizational redirection was unavoidable. ONR, against its own better judgment as to what is best for the Navy, has had to become an applied research agency focused on the Navy's short-term needs, and little else.

[93] See George H. Daniels, "The Pure-Science Idea and Democratic Culture," *Science* 156 (30 June 1967): 1699–1705.

CHAPTER FIVE

Managing Naval Science

THE HARDER TASK in government is not identifying goals, but achieving them. ONR learned through experience that the only politically acceptable objective for its research activities—the only objective that could survive scrutiny within and outside the Navy—was the enhancement of naval capabilities.[1] Some within the organization would argue that the relevance of the research that it sponsored to naval missions had come to be too immediate to serve effectively the Navy's and thus the nation's long term interests, but none doubted the necessity and appropriateness of directing research toward furthering naval capabilities. ONR's basic management problem has been to discover effective mechanisms to achieve that objective.

The problem is an extremely difficult one. To begin with, ONR must find ways to link together two complex and disparate communities. One is the Department of the Navy,[2] which since 1950 has been composed of at least one million (and at times many more) officers, enlisted personnel, and civilians who are arranged in numerous hierarchies, all of which seemingly are extremely sensitive to jurisdictional intrusions. The other is the scientific community, whose members compete for recognition, seldom agreeing upon anything except their desire for autonomy and sufficient financial support. Attempts to bridge the communities can be viewed as provocative by both. For example, the unsolicited suggestion by ONR that a naval organization might benefit from the assistance of outside scientists can be interpreted as a criticism of that organization's performance of its duties, as well as an effort to control the scope of scientific inquiry.[3]

---

[1] An internal report prepared in 1965 expresses the recognition this way: "ONR better have another justification than 'what is good for science is good for the Navy.' While this is undoubtedly true, and it is true that it has provided a serviceable assumption in the past—particularly in the more distant past when ONR was in effect the Office of National Research—it is extremely doubtful if it will provide adequate armor against the attacks which may be mounted in the foreseeable future on ONR's mission and its share of the Navy budget." *Information Flow at the Office of Naval Research*, Phase 1 Report, ACR/NAM-3, April 1965, V-4.

[2] The Department of the Navy is the civilian organizational home of two military forces: the United States Navy and the United States Marine Corps. The word naval is used within the Department to refer to matters of common interest to both the Navy and the Marine Corps.

[3] Interview document.

The agency's task is made no easier by the fact that the theoretical understanding of the innovation process is primitive.[4] We are not sure whether innovation is stimulated or stifled by resource constraints. We do not know whether it is best to set focused goals for research or to wait patiently for the unexpected. Inevitably ONR's own performance is judged by the innovations it is thought to foster in naval technologies. And yet, there is no set of policies to adopt that is certain to deliver these innovations. Instead, there are only contradictory prescriptions offered by scientists and others who claim the innovation process as a field of expertise often without the benefit of systematic investigation.

Success can never be ONR's alone. Research is but one stage in the innovation process, and not always a necessary one at that. Other agencies provide the development, testing, and operational experience required to demonstrate the effectiveness of naval equipment and concepts. Not surprisingly, they tend to assign prominence to their own contributions and express skepticism about ONR's when recognized advances occur. Given the multiple sources of support that exist for research, and the strange paths by which technical ideas are generated, the significance of ONR's contribution is easily, if not fairly, challenged.

But even when ONR is judged to have played a key role in the strengthening of a naval capability, it cannot rest. The value of past innovation diminishes rapidly in the discussions of future budgets, if for no other reason than there is constant change among the senior officials in government who must approve those budgets. It is not the past that they wish to hear about. They want achievements for which they too can claim credit. As a consequence, ONR must search continuously for projects that promise significant advances in naval technologies, projects that justify its budget allocations, and therefore for management systems that effectively link research to naval needs.

THE MANY NAVAL RESEARCH ACCOMPLISHMENTS

There is no doubt that ONR can assemble impressive listings of technical advances relevant to naval operations that have resulted from research it has supported. In fact, such lists were prepared periodically for submission to higher authorities in the hope that the selection this time would convince the skeptics among them of not only the

[4] The state of research on innovation is summarized in many places, including the Division of Industrial Science and Technological Innovation, National Science Foundation, "Program Announcement for Extramural Research: Innovation Processes Research," Washington, D.C., April 1982.

value of the organization's work, but also the appropriateness of its methods.

Some lists would stress the results of basic research.[5] The support of basic studies of interesting physical phenomena, ONR has argued, can have significant military payoffs that are initially difficult to perceive. Thus, it would be noted that fundamental work initiated in the 1940s on shock tubes and shock dynamics subsequently aided the accelerated development of reentry vehicles for ballistic missiles, or that investigations of atomic time begun before the space era led eventually to significant improvements in navigation systems, including eventually the NAVSTAR global positioning satellite system.

Other lists would stress the need to nurture certain fields that should be of special interest to the Navy because of its dependence upon them, underlying another ONR belief. Here one would learn about the advances in oceanography, meteorology, electronics, metallurgy, and aeronautics. For example, ONR would report its pioneering work in underwater acoustics so necessary for effective submarine operations and control. Or it would note that it helped support the development of the first digital computer (Project Whirlwind at M.I.T.), vital to the creation of the DEW line warning system, or the first studies of the thermochemistry of titanium and molybdenum alloys useful in the construction of missiles and high performance aircraft.

Still other lists might note needs of the Navy that require separate research efforts in fields thought to be well covered elsewhere in government. For example, the isolation of most naval units from major medical facilities necessitates special medical capabilities. Thus, ONR supports biomedical research even though NIH maintains a much larger biomedical research program. Among the achievements ONR cites are pioneering work in the freeze storage of whole blood (valuable in the treatment of wounds aboard ship), new techniques for the care of burn victims, and the development of topical fluoride compounds for the dental hygiene of submarine crews.

When the concern is thought to be the development of tactical fleet support systems, ONR has another set of significant accomplishments to report, mostly those produced by its applied research program.[6] It initiated the early design of integrated air defense computer networks that led to the development of the Naval Tactical Data System widely used in the U.S. and NATO navies, and it promoted the concept of

[5] See, for example, Appendix E, Review of Navy R&D Management, 1946–1973.

[6] *Naval Research Utility*, vol. 2, Appendix D. Also an address by Hon. Garrison Norton, Assistant Secretary of the Navy for Air, *A Decade of Basic and Applied Science in the Navy*, a symposium sponsored by ONR as part of its decennial year, 19–20 March 1957 (Washington, D.C.: Government Printing Office, 1957).

towed arrays for submarine detection. The latter work, which has added greatly to antisubmarine warfare capabilities, was championed nearly single handedly by Marvin Lasky, an ONR scientist who received the Defense Department's Distinguished Civilian Service Award for his effort. In addition, ONR supported work on various computerized training aids to reduce instructional time for the training of technical specialists and aided the advanced development of logistical systems for the on-time supply of crucial spare parts to the fleet.

The Office also believes that its support of research, both basic and applied, provides a useful reservoir of expertise and equipment upon which the Navy can draw during emergencies. One example often cited is that of deep diving, submersible technology. ONR began supporting the construction of deep diving research vehicles in the late 1950s. When the Navy needed the assistance of a deep diving submersible in 1966 to recover a nuclear weapon lost off the coast of Spain in an aircraft accident, ONR had the oceanographic research vehicle ALVIN available. The quick recovery of the bomb helped reduce the political damage to U.S. interests in Spain and elsewhere caused by the incident.[7]

Assembled as one, the lists describe the modernization of the Navy in the years since the Second World War. During this period naval strength has become ever more dependent upon the advance of technology. The ships and aircraft of the fleet bristle with sophisticated target acquisition sensors; their crews man the most exotic weapons. And while the application of this naval force is not risk-free, neither is its development. Without the courage of its convictions, ONR could not have contributed as much as it has to the defense of the nation.

And a Singular Failure

The lists of significant contributions notwithstanding, some officials (including successive Chiefs of Naval Research) came to believe that ONR's performance could be improved. They were not persuaded that past innovations justified current budgets. Instead, they wanted to be assured that current resource allocations would produce important future improvements in naval capabilities. Their desire was for the institution of a management system that would guarantee naval-rele-

---

[7] See Edward H. Shenton, *Diving for Science: The Story of the Deep Submersible* (New York: W. W. Norton, 1972), 66, 126, 142–43; Admiral Horacio Rivero, "Innovating in Support of Naval Operations," in F. Joachim Weyl, ed., *Research in the Service of National Purpose* (Washington, D.C.: Government Printing Office, 1966), 14–18. Commander S. Monsen won the Legion of Merit for this work. Note also Julia M. Ford, "NSRDC and the New ALVIN," *Naval Research Review* (January 1974): 9–19.

vant results of ONR's research investments. It was a desire that the ONR staff could not fulfill.

The pressure for such a management system began in the early 1960s when Robert McNamara became Secretary of Defense. The Navy in particular resisted McNamara's emphasis on management reforms, senior officers viewing it as a threat to the service's traditional independence. But some ambitious junior officers did see a logic in the reforms being forced upon the military establishment. There was, they were convinced, a need to become cost conscious in the use of defense resources. Research allocations seemed especially out of control, being made more on the whim of civilian scientists than on the basis of objectively established military requirements. It was these officers who learned the language of reform and who defended its introduction. Their understanding of the new realities of defense management caught the attention of senior civilians within the department who rewarded them with influential assignments. Rear Adm. John Leydon, appointed Chief of Naval Research in 1964, was among them; so too was his successor, Rear Adm. Thomas Owen, appointed Chief of Naval Research in 1968.[8]

Despite personal doubts about the applicability of the new management doctrines to research, ONR staff members had no choice but to assist their exploration within the Navy's research program. Ironically, some of these management concepts sprang from ONR's own pioneering work in management science that began in the 1950s. Urged on by Admiral Leydon, studies were initiated on the cost-effectiveness of research and the quantification of research requirements.[9] Reorganization alternatives were prepared and new management procedures calling for the frequent review and justification of program elements were tried.[10]

Perhaps the most ambitious of these efforts was the research matrix scheme proposed by Comdr. George Hoover, a retired naval aviator

---

[8] Luther J. Carter, "Office of Naval Research: 20 Years Bring Changes," *Science* 153 (22 July 1968): 397–400; Interview documents.

[9] For example, J. Bruce Davis, "Annotated Bibliography on Methods for Evaluating Basic Research Projects," Technical Memorandum No. 105, Operations Research Department, Case Western Reserve, April 1968; R. A. Goodman and W. J. Abernathy, "Summary of a Workshop on Dimensional Analysis for Design, Development and Research Executives," University of California at Los Angeles, October 1971. ONR staff even anticipated the pressure for management controls in an internal study, *Final Report by Ad Hoc Study Team on Research Planning*, Office of Naval Research, 15 March 1963, which called for the support of methodological studies and a better sales effort for ONR accomplishments.

[10] *Study of Organization, Functions and Staffing of the Office of Naval Research*, a report by the Chief of Naval Research to the Secretary of the Navy, August 1965.

who had become a defense management consultant.[11] The matrix juxtaposed operational requirements and research fields in an attempt to determine optimal research investments. Rather than perpetuate the pattern of incremental adjustments in allocations among fields, the matrix arrangement promised, if implemented, to alter drastically budgetary divisions. Commander Hoover's enthusiasm for the scheme won over most of the naval officers within ONR, but not many of the organization's civilian scientists, who considered it to be both overly elaborate and unnecessary.[12] Their opposition eventually scuttled the plan to make the matrix ONR's basic management tool. They offered no substitute.[13]

The conviction that ONR needed a formal management system to identify naval research opportunities and to direct its scientific activities lingered, creating a wedge within the organization. Those who sought management reform, primarily naval officers, wanted assurance that science supported by the Navy would serve significant naval objectives. Those who thought a more opportunistic approach was best (primarily civilians) believed that the future, both of warfare and of science, was too uncertain for the imposition of rigid programming systems. Each side in the ongoing dispute began to doubt the sincerity of the other, causing further deterioration of relations within the organization. At one point, the naval officers began to hold their own office meetings and considered removing all but a handful of the civilians. In turn, the civilians devised schemes to isolate military influences within the organization.[14]

A STRUCTURAL ANOMALY

Naval officers have acquired an increased interest in the management of research in large part due to a series of reorganizations which unintentionally made ONR a structural anomaly within the department. When ONR was established at the end of the Second World War the Navy had essentially the same bilinear structure it had acquired in the previous century. That is, the Navy was divided into two parts, each

[11] Commander Hoover's exciting naval career is described in Commander Ted Wilbur's article, "Let George Do It," *Naval Aviation News*, December 1971, 8–19. Matrix methodology is discussed in Marvin J. Cetron, et al., *Technical Resources Management: Quantitative Methods* (Cambridge, Mass.: M.I.T. Press, 1969), chaps. 2, 4.

[12] Interview documents.

[13] The conclusion that nothing special is needed to improve ONR performance was drawn from a major study of the applicability of research management methods conducted internally during 1967 and 1968. *Naval Research Utility*, 2 vols., ONR Memorandum DR/NAM-6, July 1968.

[14] Interview documents.

of which reported separately to the Secretary. One of these, the Shore Establishment, was composed of several separately chartered bureaus (e.g., Ships, Ordnance, Yards and Docks, etc.) and was concerned with materiel development and support. The other, the fleet, had its own set of commands and was concerned with operational issues. The war had changed this arrangement only slightly. The Chief of Naval Operations (CNO), until then only a relatively small planning office, acquired control over the fleet. The legislation creating ONR appropriately assigned the organization to Shore Establishment and gave it bureau status, including a direct reporting line to the Secretary. The CNO, whose jurisdiction was at the time confined to fleet matters, had no authority over ONR or other naval research activities.

The creation of the Department of Defense in 1948 began a process of bureaucratic centralization in defense affairs that eventually altered these relationships. Largely as a result of the Navy's opposition to centralization, the Department of Defense's power over the service was at first quite limited.[15] The political frustrations caused by the inability to control increasing defense expenditures, however, led to several congressionally mandated reorganizations of the department. The most important of these occurred in 1958 when the Secretary of Defense, through the Joint Chiefs of Staff, was given command over all operational forces, including deployed naval forces. Through the Director of Defense Research and Engineering, he was also given responsibility for the development and procurement of military equipment and supplies. The most immediate effect of the 1958 reorganization was to eliminate the direct control of the CNO's staff over the fleet, though the CNO himself was to continue to be involved in the chain of command through his membership in the Joint Chiefs of Staff, which is composed of the senior officer of each of the services. Also eliminated, although not initially recognized, was the Navy's ability to maintain its bilinear structure.

Secretary McNamara forced this recognition when he insisted in the 1960s that the Army and the Navy adopt an internal format similar to that of the Air Force, which had previously consolidated all of its materiel support and acquisition activities into a single command subordinated to the Air Staff. The quest for centralized control over weapon acquisition within the Navy was resisted by several bureau chiefs, who managed only to achieve a brief delay and their own early retirement. In 1966 the bureaus, renamed systems commands, were

[15] Demetrios Caraley, *The Politics of Military Unification* (New York: Columbia University Press, 1966); Albion and Connery, *Forrestal and the Navy*; Vincent Davis, *Postwar Defense Policy and the U.S. Navy, 1943–1946* (Chapel Hill: University of North Carolina Press, 1962).

made elements of the newly-formed Naval Materiel Command, which reported to the CNO rather than to the Secretary of the Navy.[16]

Prior to the establishment of the Naval Materiel Command, OPNAV had held responsibility only for the identification of naval R&D requirements. It was the responsibility of the various Navy technical units, including the bureaus and ONR, to develop equipments and techniques to meet these requirements. With the 1966 reorganization, OPNAV was assigned responsibility for managing nearly all of the Navy's R&D effort, as well as directing weapon procurement activities. ONR's research programs were formally exempt from OPNAV control as ONR, tied by statute to the Secretary of the Navy, was not made part of the Naval Materiel Command.

Nevertheless, ONR could not help but be affected by the reorganization. With the fleet out of their grasp, many of the senior staff admirals of the Navy—the admirals who populate OPNAV—found fulfillment in the management of the Navy's weapon procurement and R&D efforts. ONR's special status offered little protection from a flotilla of admirals bent on demonstrating their individual prowess in overseeing the Navy's vast technical enterprise. Each seemingly had a favorite doctrine to explore, a favored project to promote. And few among them appeared aware that they might have something to learn from those with experience in directing research and development activities.

The justification for the consolidation of naval R&D activities into a single command and its subordination to the CNO was that it would improve coordination within the Navy. The Navy's various technical agencies, it was argued, had overlapping jurisdictions, duplicated each other's efforts and were unresponsive to operational needs because of their independence. Consolidation of the R&D effort would eliminate wasteful rivalry. Having the Naval Materiel Command report to the CNO would link R&D planning more effectively to warfare planning.

However, many civilians within ONR believed the reorganization was a serious mistake. Centralization, they thought, would stifle the

---

[16] Booz Allen & Hamilton, Inc., *Review of Navy R&D Management 1946–1973*, report prepared 1 June 1976, Summary Volume, 11; Scott MacDonald, "How the Decisions Were Made: Exclusive, Inside Story of Navy Reorganization," *Armed Forces Management*, May 1966, 74–79; Capt. Thomas McGrath, USN (Ret.), "CNM—'Super Bureau' in Navy Reorganization," *DATA*, May 1966, 20–21; "Navy Reorganizes Materiel Command Structure," *Defense Industry Bulletin*, April 1966, 12–13; Order of the Secretary of Navy to ALNAV, subject: Activation of the Naval Materiel Command, 29 April 1966. The Naval Materiel Command was abolished in a 1986 reorganization initiated by Secretary of the Navy John Lehman.

creativity and flexibility historically characteristic of naval research and development activities. Rather than improve coordination, the reorganization would hinder it by adding several levels of bureaucracy to the decision-making process. Worse yet, they feared that although ONR stood formally apart from the new command structure, its officers, if they valued their careers, would seek direction for managing research from the admirals in OPNAV rather than continue ONR's pattern of balancing the interests of science and the Navy.

Constrained Autonomy

An underdeveloped field though it may be, the sociology of science offers insight into an effective way to organize scientific research. Studies by Pelz and others indicate that research groups are most productive when they have some security, but not total security.[17] That is, groups whose financial and organizational situation is guaranteed in the short run, but not necessarily for the long run, perform better than do groups that either lack any guarantees or who have permanent guarantees. The groups without some assurance of support are too involved in seeking financial backing to be productive, while those with total security find other things to do instead of work. An argument perhaps against academic tenure and for three- to four-year contracts.

Parallel conditions would appear to exist for agencies supporting applied science. These organizations, no less than the scientists that they support, are affected by environmental uncertainties. Sanford Weiner has argued that agencies that had some independence from operational concerns, but not total isolation, were more effective in serving governmental objectives than were agencies that lacked that independence or that were totally free from any links to operational activities.[18] Writing in the early 1970s, he identified ONR as an agency whose structural arrangements produced the effective combination of short-term independence and long-term uncertainty, an organizational condition that Weiner labeled as constrained autonomy.

The Office's civilian leadership at the time would certainly have agreed with this assessment. They believed that ONR's placement within the Navy's Secretariat had distanced the organization, at least until the late 1960s, from the immediate political interests of the Navy—interests aggregated and defended by OPNAV. They were

[17] Donald C. Pelz and Frank M. Andrews, *Scientists in Organizations* (New York: John Wiley & Sons, Inc., 1966).

[18] Sanford L. Weiner, "Resource Allocation in Basic Research and Organizational Design," *Public Policy* 20, no. 2 (spring 1972): 227–55.

aware, however, that the organization's long-term survival had, since the Korean War, come to depend upon the rest of the Navy being convinced that ONR was making an effective contribution. There was sufficient tension in ONR's situation to keep it productive, but not enough to paralyze it.

The contrast with other government research organizations is instructive. For example, the Air Force's Office of Scientific Research (AFOSR) never gained the stature that ONR had within and outside its own service, due in large part apparently to its lack of organizational autonomy.[19] When the Air Force was formed from the Army Air Force after the end of the Second World War, it adopted much of the organizational format of its parent service, which had acquired a tradition of having a strong military staff and a relatively weak civilian secretariat. Research in the Air Force was always under the purview of the Air Staff and thus subject to the whims of the military.

The origins of AFOSR can be traced to the establishment in 1948 of an Office of Air Research in the Air Material Command, then the Air Force's weapon development and procurement agency.[20] The ability of the Office of Air Research to carry out its mission—the conduct and support of basic research relevant to Air Force interests—was greatly hindered by the procurement mentality that dominated the Command. Air Material Command contracts officers, for instance, saw no need to treat basic research contracts with universities any differently from aircraft parts contracts with industrial firms. Academic research, in their view, was just another commodity to be purchased.[21]

In 1951 most of the Air Force's R&D activities, including the Office of Air Research, were pared off from the Air Material Command to form a new command in hopes that they would be able to flourish with more independence. Although this was undoubtedly the case for most of the units transferred, it was not for the Office of Air Research. During the ten years it was part of the Air Research and Development Command (ARDC), the Office of Air Research, renamed the Office of Scientific Research, found itself locked in jurisdictional and budgetary struggles.[22] The ARDC's in-house laboratories wanted authority to

[19] Ibid., 238–42.

[20] "History of AFOSR," *AFOSR Research*, Air Force Office of Scientific Research, Office of Aerospace Research, U.S.A.F., Arlington, Va., July 1967, 12.

[21] The best source of information on the AFOSR experience during this period is the candid official history prepared by Nick A. Komons, *Science and the Air Force* (Arlington, Va.: Office of Aerospace Research, 1966). Also Robert Sigethy, "The Air Force Organization for Basic Research 1945–1970," Ph.D. dissertation, American University, 1980.

[22] Komons, *Science and the Air Force*, chaps. 2, 7.

conduct their own university research contract programs. Given that ARDC's prime interest was weapons development, however, there was disinclination on the part of its senior officers to fight very hard for basic research allocations in their budget requests. At one time, OSR's funds were hidden in appropriations for B-52 wing modifications. On another occasion it faced near elimination by a barely-averted budget slash. As one ARDC Commanding General described the situation, "I would never be criticized [by the Air Staff] for what I didn't do in [basic] research."[23]

When the ARDC and the Air Material Command were combined into the Air Force Systems Command in 1961, OSR and other research activities were transferred to the Headquarters of the Air Force (the Air Staff). It was obvious to all that research of any type would be ignored in an organization established to improve the coordination of weapon acquisition activities. Initially, OSR was reinvigorated by its apparent increase in status. It was, thanks to the reorganization, only one rung below the Chief of Staff. But when the services were pressured in the late 1960s to demonstrate again the relevance of their research, OSR, as a component of the Air Force's headquarters, was in a very exposed position.[24]

The National Institutes of Health (NIH), until recently at least, has had a quite different experience, one best characterized as organizational autonomy without constraint.[25] Begun in 1937 with a single institute (the National Cancer Institute) and a modest appropriation of a few hundred thousand dollars, NIH now has a dozen institutes and an annual budget approaching $8 billion. Its period of most rapid growth occurred during the 1950s and early 1960s when the support of biomedical research became identified as the most acceptable alternative politically to national health insurance and other direct financing schemes for expressing the federal government's concern about

[23] Lt. Gen. Thomas S. Power quoted in "Air Force Boosts Status of Research," *Aviation Week* 64 (6 August 1956): 326.

[24] The Air Force quickly eliminated 10 percent of OSR's projects after their relevance to Air Force activities was questioned by the General Accounting Office. It also assured Congress that it would keep closer control over OSR projects. In contrast, few projects were eliminated from ONR's roster as the Secretary of the Navy came to ONR's defense. U.S. General Accounting Office, "Need to Strengthen Management Control Over the Basic Research Program Administered by the Air Force Office of Scientific Research," 29 January 1971. Interview documents.

[25] Stephen Strickland, *Science and Dread Disease* (Cambridge, Mass.: Harvard University Press, 1972); James A. Shannon, "Federal Support of Biomedical Sciences: Development and Academic Impact," *Journal of Medical Education* 51, no. 7, part 2 (July 1976): 1–97; Stephen Strickland, *Research and the Health of Americans* (Lexington, Mass.: Lexington Books, 1978).

the nation's health. A powerful coalition of congressional champions (headed by Senator Lester Hill and Congressman John Fogarty) and lobbyists (including U.S. medical schools, various professional societies, voluntary health associations, and one very determined philanthropist and political campaign contributor, Mary E. Lasker) helped the then director of the institutes, Dr. James Shannon (himself a highly skilled bureaucratic politician), to create the world's premier biomedical grants agency.[26]

Biomedical scientists could not have been more satisfied than they were with the administrative mechanism NIH selected to distribute its largess. Even though it received appropriations on the basis of disease categories (e.g., cancer, heart, dental), NIH chose to allocate its grants through what it called study sections, disciplinary-based groups (e.g., molecular biology, neurophysiology) composed almost entirely of nongovernment scientists. Periodically, the study sections would convene as peer review panels to judge applications for support, ranking projects solely on the basis of their perceived scientific merit. NIH administrators would then award grants according to the study section rankings to the point at which the funds would be exhausted. The tally of unfunded but merit-approved projects, purposely kept high, would be used in budgetary requests to indicate a need to increase appropriations. Given that NIH administrators were largely drawn from (and would most likely retire to) the institutions at which the grantees worked, it was not surprising that their priorities and values would be nearly identical with those dominant within the biomedical research community.

By the end of the 1960s, however, the coalition that supported NIH's growth began to break apart. The politically sensitive members of the coalition (Mary Lasker among them), felt compelled to insist that NIH demonstrate its contribution to improvements in the nation's health status rather than its contribution to advances in basic science. Less friendly forces questioned whether NIH's contribution could be anything other than increases in health care costs, now partly paid through governmental programs such as Medicare and Medicaid. Pressure developed to direct more of NIH's research toward specific goals and to alter its management structure to increase the agency's public accountability.[27]

[26] Strickland, *Research and the Health of Americans*; also Elizabeth Drew, "The Health Syndicate," *Atlantic Monthly*, December 1967, 75–82.

[27] The pressure for reform began with the "Wooldridge Report," *Biomedical Science and its Administration: A Study of the National Institutes of Health*, Report to the President, February 1965. See also Joseph D. Cooper, "Onward the Management of Science: The Wooldridge Report," *Science* 148 (11 June 1965): 1433–39; John Walsh,

The attempts to restrict NIH's autonomy have been only partially successful. The desire for goals produced a war on cancer followed by dozens of other earmarked research efforts. Resources have increased, but so too have searching questions. More recently, the NIH budget growth has slowed, and much more of the allocations are absorbed by unwanted applied research tasks. But the biomedical research community still holds strongly to the belief that the support of basic research should be at the center of NIH's attention, and still very much influences the agency's administration. The political struggle to control the direction of the nation's biomedical research effort continues.

PROGRAM ENTREPRENEURS

The extent to which ONR actually took advantage of the independent status it held within the Navy depended largely upon the initiative of its program managers, the civilian scientists who directed the contract research effort. Naval officers could cajole and complain, but with rare exceptions they lacked the scientific knowledge needed to identify promising areas of research and the truly productive researchers. Scientists with important ideas rarely began their work with sufficient knowledge of the Navy to discover vital applications for these ideas. When science and the Navy meshed within ONR, it was usually because of the willingness and ability of a program manager to broker the relationship.

Unlike NIH and NSF, ONR did not rely very much on peer review panels to select the research projects it would support. The decision to use program managers was made when the contract research program was established in 1946. Later, the rationale would be offered that as a military contract agency, ONR could not let outside scientists guide its research judgments.[28] Conveniently ignored was the fact that both the Army and the Air Force used peer review panels (usually established under contract with the National Research Council and similar bodies) extensively to guide their decisions, and the fact that ONR itself used peer review panels for most of its biomedical research

---

"NIH: Demand Increases for Applications of Research," *Science* 153 (8 July 1966): 149–52; "LBJ at NIH: President Offers Kind Words for Basic Research," *Science*, 157 (28 July 1967): 403–5; John K. Iglehart, "The Cost and Regulation of Medical Technology's Future Policy Directions," *Milbank Memorial Fund Quarterly* 55 (winter 1977): 61–79; Richard Rettig, *The Cancer Crusade* (Princeton, N.J.: Princeton University Press, 1978); Harvey M. Sapolsky, "Here Comes the Artificial Heart," *The Sciences*, December 1978, 25–27.

[28] Interview document.

programs (apparently due to the intense preference among biomedical scientists for this system),[29] and occasionally when starting a new program.[30] More likely, Captain Conrad made the correct assessment that the universities that he was seeking to involve in the contracts program would not feel comfortable cooperating, and would not hold much respect for ONR, unless he had a staff of first-rate scientists to administer the program. These scientists, in turn, would not have agreed to stay with the contracts program unless they were given the opportunity to exercise their own judgment as to the direction of the research. Thus did ONR manage to avoid much of the vagaries of one project selection method to acquire those of another.

Critics have argued that the peer review system, because of its tendency toward consensus, is inherently conservative, favoring established investigators following the standard paradigms of a discipline and who are affiliated with prestigious institutions.[31] It is often attacked as an old boy network in which favors are traded among the favored. There is also a concern among scientists that the members of peer review panels may appropriate the ideas revealed in the proposals. Some even admit to submitting only completed work and to using any awards for projects not revealed in order to avoid being preempted by the scientists serving on the panels.[32] Were it not for the fear that Congress would mandate geographic quotas or some other more arbitrary arrangements for distributing research awards, more scientists might speak out against the peer review system. And then there is the criticism that peer review abdicates government accountability. With peer review, agencies do not make choices; peers do. But the demo-

[29] Interview document. One of the more interesting mysteries in the sociology of science is the origins of the peer review system and especially its absorption as a norm within the biomedical research community. The Rockefeller Fund used peer review for the medical grants it distributed in the 1930s, but why the method was selected is unknown. Few biomedical scientists are aware of the fact that the NIH did not begin life totally committed to peer review. Its practice at first was to have double screening of peer panels and in-house government scientists for all external grants. The opposition to the role of government scientists came from academic scientists who feared competition from government laboratories. Interview document.

[30] Interview document.

[31] See Thane Gustafson, "The Controversy Over Peer Review," Science 190 (12 December 1975): 1060–66; John Walsh, "NSF Peer Review Hearings: House Panel Starts With Critics," Science 189 (8 August 1975): 435–37; William D. Carey, "Peer Review Revisited," Science 189 (1 August 1975): 331; "NAS President Lauds NSF Peer Review System," Chemical and Engineering News, 7 March 1977, 14–15; National Institutes of Health, "Decisions by Director NIH on Recommendations of Grants Peer Review Study Team," 8 February 1978.

[32] Interview documents. Also Charles Kidd, American Universities and Federal Research (Cambridge, Mass.: Harvard University Press, 1959), 199.

cratic control of administration requires agencies, and ultimately elected officials, to take responsibility for the use of public money rather than to hand over their responsibility totally to private, self-selected groups.

The program manager system was not without its own flaws, and very human ones at that. Some ONR program managers were suspected of supporting the work of certain prominent scientists only to enhance their own standing within their disciplines. Others were thought to have an overly strong attachment to their alma maters. And once, a directive had to be issued urging program managers not to support research in Europe just so to have an official reason to travel there at government expense.[33]

But the system's weakness was also its strength. A program manager built on his own program. Each had a certain scientific or warfare jurisdiction, often poorly defined, a limited amount of money, and a license to search for technological opportunities. Some preferred the conventional, while others took risks. Some floundered, while others did not. The best contributed significantly to the technical prowess of the Navy and the nation.

There were, to be sure, some unofficial guidelines. Program managers quickly learned that certain research topics, regardless of their scientific interest or relevance to naval activities, were best avoided. For example, because of the controversy surrounding Project Camelot (an Army-sponsored study of the sources of societal instability in the Third World that was terminated after it was attacked as imperialistic), foreign area behavioral science investigations were subject to such time-consuming clearance procedures at the State Department and Department of Defense, that few were supported.[34] Similarly, as opposition to the war in Vietnam became an irritant to the administration, scientists who took an active role in mobilizing opinion against the war were quietly dropped from the rolls.[35] As one program manager put it, knowing who and what was politically dangerous to support was an important part of the program manager's job.[36]

But the politics that truly mattered were the politics of the Navy, not national politics. An accurate vision of the Navy is that of an elaborate bureaucratic maze, a puzzlelike web of commands, offices, laboratories, and activities. In order to be effective, program managers had to learn to negotiate their way through its intricacies; they had to

---

[33] Interview documents.

[34] Interview document. A useful analysis of the Project Camelot controversy is Robert A. Nisbet, "Project Camelot: An Autopsy," in Philip Rieff, ed., *On Intellectuals* (Garden City, N.Y.: Doubleday, 1969), 283–95.

[35] Interview documents.

[36] Interview document.

discover the ties of friendship, professional identities, and technological interests that outline power relationships within the Navy and that could describe a possible safe path for whatever ideas they wished to champion.[37]

The challenge was always to translate the theoretical into the practical; scientific concepts into potential military systems, procedures, or equipment.[38] Good ideas, however, were never sufficient. The fit had to be found within the existing organizational framework of the Navy. Here a knowledge of the cleavages that divide the Navy could be used to advantage. Perhaps the aviators needed help in their struggle with the ship drivers or the missileers in theirs with the gun club. Would the destroyer advocates find the concept useful or would it be the helicopter advocates? There could be no success until some group within the Navy was persuaded to adopt and defend the project.

Acquiring well-placed allies within the Navy, in turn, helped secure a base for the manager's program within ONR. Such allies could provide letters of endorsement or even additional funds, the existence of which was taken to be certification of the program's naval relevance.[39] They could always ward off the designs of ONR budget planners. For example, neither amphibious warfare nor geographic science ranked high among ONR's priorities. But when the geography branch decided to emphasize coastal zone science, it gained the continuing support of the Marine Corps, whose commitment to amphibious warfare and concern about beaches appears permanent. And because the Navy frequently used the Marine Corps to advocate favored aircraft and ship projects before Congress, the Marine Corps' small wishes within the Department had to be granted.[40]

Most often though, the support gathered was fleeting. During the years it might take to nurture a scientific advance, the Navy's political alignments do not remain constant. Personnel transfers, retirements, and changing doctrines within the Navy forced the program managers to search continually for allies.[41] Only the most tenacious program managers could withstand the frustrations inherent in achieving a major innovation. The majority settled for less.

The agency's program was the summation of usually small-scale re-

[37] As one analysis correctly perceived, "There is a lack of uniform mission oriented guidance in a form which is appropriate for the disciplinary oriented resource allocation decision in ONR. As a result each decision-maker must independently seek out guidance sources, analyze them, and translate them into guidance information relevant to his decision making needs." Abt Associates, "A Methodological Study of Mission Oriented Basic Research Planning in the Office of Naval Research," March 1965, 12.

[38] Interview documents.

[39] Interview documents.

[40] Interview documents.

[41] Interview document.

search projects nurtured by its civilian scientists and the occasional naval officer turned research zealot. At times, by accident or assignment, the organization would get involved in a much larger undertaking. Failure was the frequent, although not inevitable, outcome. In the late 1950s the Chief of Naval Research (by virtue of his supervision of the Naval Research Laboratory) was placed, if only indirectly, in charge of the Vanguard Project, the ill-fated attempt to launch America's first space satellite.[42] (As much as ONR wanted to distance itself from the project once Sputnik was launched, it could not. The Chief of Naval Research at the time, Rear Adm. Rawson Bennett, became the Navy's Vanguard spokesman and quickly put his foot very deep into the muck of politics as the *New York Times* captured with this headline: "Admiral Says Almost Anybody Could Launch 'Hunk of Iron.'"[43] Two months later, the "more sophisticated" Vanguard burned on its launch pad before the assembled media of the world.) Somewhat later, ONR became involved (again through the Naval Research Laboratory) in the construction of a giant radio telescope at Sugar Grove, West Virginia, a project that was terminated after amassing considerable cost overruns.[44] And for over twenty-five years ONR supported high altitude balloon studies including a series of manned flights that ended after the death of one of the pilots.[45] Although some important new knowledge was gained in each case, the national security interest promoted for the project seemed to take it beyond ONR's scientific and management capabilities.

What ONR did best was to give its science administrators the freedom to broker relationships between scientific ideas and naval applications.[46] No management system could predict which ones would succeed. But some would, as long as the leadership of the agency could fend off the pressures to guarantee that none would ever fail.

[42] For a detailed analysis, see Constance McLaughlin Green and Milton Lomask, *Vanguard—A History* (Washington, D.C.: National Aeronautics and Space Administration, 1970).

[43] *New York Times*, 5 October 1958, 2.

[44] Daniel Greenberg, "Big Dish: How Haste and Secrecy Helped Navy Waste $63 Million in Race to Build Huge Telescope," *Science* 144 (29 May 1964): 1111–12; General Accounting Office, *Unnecessary Costs Incurred for Naval Radio Research Station Project at Sugar Grove, West Virginia*, April 1964; Frederic H. Jacobs, "Radio Astronomy and the Navy: Buildup to Breakdown," unpublished paper, M.I.T., May 1971.

[45] Commander Ed Melton, USN, "Twenty-Five Years of SKYHOOK," *Naval Research Reviews*, February 1973, 1–12; Interview documents. The scientific aspects of naval balloon research are described in Henry Remaboski, "The Role of High Altitude Balloons in Space Research," paper presented at the AIAA/NASA/VARC Seminar for University Research Using Sounding Rockets, Balloons, and Satellites, 1 March 1967, at Williamsburg, Virginia.

[46] Interview documents.

CHAPTER SIX

Science Advice for the Navy

PUBLIC LAW 79–588, the statute which established ONR in August 1946, also authorized the Secretary of the Navy to create a Naval Research Advisory Committee (NRAC), consisting of not more than fifteen preeminent civilian scientists, to advise the Chief of Naval Operations and the Chief of Naval Research on matters relating to research. The committee was formed the following October and included among its initial members such well-known figures in science as Karl Compton, the President of M.I.T.; his brother Arthur Compton, a physicist at the University of Chicago; Detlev Bronk, the President of the National Academy of Sciences; and Warren Weaver, Director for Science at the Rockefeller Foundation.

Similar committees, with overlapping memberships, were subsequently established for the other armed services,[1] the Department of Defense, most major agencies in the federal government, and, after Sputnik, for the President of the United States.[2] The existence of this network of interlocking advisory committees was taken by most observers to indicate the ascendancy of a scientific elite into the center of national policy making. A relatively small number of scientists and engineers drawn from a few prestigious institutions were thought to hold great influence in the councils of government, affecting decisions on such important issues as the allocation of budgets and the formulation of national strategy.

A literature, modest to be sure, developed in the early 1960s to ex-

[1] The Air Force, tracing the origins of its advisory committee to the days of the Army Air Force, would claim to have the first of the modern science advisory committees. See Thomas A. Sturm, *The USAF Scientific Advisory Board—Its First Twenty Years 1944–1964* (Washington, D.C.: Government Printing Office, 1968).

[2] The establishment and role of the President's Science Advisory Committee is discussed in Eugene B. Skolnikoff and Harvey Brooks, "Science Advice in the White House? Continuation of a Debate," *Science* 187 (1975): 35–37. As they point out, a predecessor committee existed in the Office of Defense Mobilization beginning in 1951. See Detlev W. Bronk, "Science Advice in the White House: The Genesis of the President's Science Advisors and the National Science Foundation," *Science* 186 (11 October 1974): 116–21. See also H. Guyford Stever, "Science Advice—Out Of and Back Into the White House," *Technology in Society* 2 (1980): 61–75, and I. I. Rabi, "The President and His Society 2 (1980): 61–75, and I. I. Rabi, "The President and His Scientific Advisers," *Technology in Society* 2 (1980): 15–26.

amine the scientist's role as adviser.[3] Given a general presumption that science advisers were influential in government, discussion tended to focus on the desirability of scientists affecting policy. Some commentators warned of the danger of important decisions being made on the secret counsel of a few individuals unrepresentative of, and unaccountable to, their professions and society.[4] Most, however, greeted with enthusiasm the apparent acceptance of scientists within the inner circles. Government decisions would be more rational, it was said, as there would now be the benefits of expert advice on the technical components of public problems. Moreover, scientists were thought to possess unique characteristics that qualified them for special access to the policy making process.[5] They were objective by training, independent in status, and optimistic in orientation. They were also supposedly better able to predict the future, it was argued, because the future "lay in their bones."[6]

By the late 1960s, however, there was considerably less discussion of scientists as an influential elite. President Johnson had been shamefully open in ignoring his science advisers; President Nixon fired his. Although much of the rest of the advisory apparatus remained intact, the scientists manning it appeared confined to the periphery of policy making. No longer was there talk of science czars and of professors shuttling back and forth between Cambridge or Berkeley and Washington. Rather there was only a growing lament within the scientific community that it was not being consulted on the major policy issues.[7]

Rare in the literature on science advice is an analysis of the conditions in which such advice will be sought and the motivations of sci-

[3] See the essays in Robert Gilpin and Christopher Wright, eds., *Scientists and National Policy Making* (New York: Columbia University Press, 1964) passim, especially those by Robert C. Wood and Don K. Price; Daniel S. Greenberg, "The Myth of the Scientific Elite," *Public Interest* (fall 1965): 61–70; Donald Cox, *America's New Policy Makers: The Scientists' Rise to Power* (New York: Chilton, 1964); Ralph Lapp, *The New Priesthood: The Scientific Elite and the Use of Power* (New York: Harper and Row, 1965); Avery Leiserson, "Scientists and the Policy Process," *American Political Science Review* (summer 1965): 408–16.

[4] Lapp, *New Priesthood*, represents this view.

[5] See, for example, essays by Robert Wood and Warner Schilling in Gilpin and Wright, *Scientists and National Policy Making*.

[6] C. P. Snow, *Science and Government* (Cambridge, Mass.: Harvard University Press, 1961) 72.

[7] A common theme in articles in *Science* during the late 1960s and early 1970s. See, for example, Phillip M. Boffey, "Hornig Years: Did LBJ Neglect His Science Advisor?" *Science* 159 (31 January 1969): 532–34, and Phillip M. Boffey, "NSF Directorship: Why Did Nixon Veto Franklin A. Long?" *Science* 164 (25 April 1969): 406–12. Note also David Z. Beckler, "The Precarious Life of Science in the White House," *Daedalus* 103 (1974): 115–34.

entists in offering it. Absent is an appreciation of the bureaucratic uses of advice and the mechanisms available to control the influence of outside advisers in matters of vital interest to agencies. This is so because the assessment of the science adviser's role in government has been derived largely from the anecdotes and autobiographies of the advisers themselves rather than from independent analyses.[8] There are few studies of the activities of particular committees that are neither official accounts nor reports of participants.[9]

Here I examine NRAC's experience utilizing, among other sources, the record of its early meetings. NRAC was not the only route by which the Navy obtained the advice of scientists during this period, but as the senior departmental advisory committee, it was kept generally appraised of the activities of other Navy science advisory committees and attempts by the department officials to solicit advice from scientists on policy issues. Thus, there is also the opportunity to explore the reception afforded science advice within the Navy during the years when the influence of scientists was supposedly at its peak.

[8] Among the more prominent autobiographies are those of Vannevar Bush, *Pieces of the Action* (New York: Morrow, 1970); Theodore von Karman with Lee Edson, *The Wind and Beyond* (Boston: Little, Brown & Co., 1967); Warren Weaver, *Scene of Change* (New York: Scribner, 1970); George B. Kistiakowsky, *A Scientist in the White House* (Cambridge, Mass.: Harvard University Press, 1976); and James R. Killian, *Sputnik, Scientists, and Eisenhower* (Cambridge, Mass.: M.I.T. Press, 1977). A much more analytical view of the science advisory process by a leading participant is Harvey Brooks' essay on science advice in his *Government of Science* (Cambridge, Mass.: M.I.T. Press, 1968). Note also Edward J. Burger, Jr., *Science at the White House* (Baltimore: Johns Hopkins University Press, 1980) and Skolnikoff and Brooks, "Science Advice in the White House? Continuation of a Debate."

[9] Important exceptions are Robert Gilpin, *American Scientists and Nuclear Weapons Policy* (Princeton: Princeton University Press, 1962); Bruce L. R. Smith, *The RAND Corporation* (Cambridge, Mass.: Harvard University Press, 1966); and Thomas A. Sturm, *The USAF Scientific Advisory Board—Its First Twenty Years 1944–1964* (Washington, D.C.: Government Printing Office, 1968). The general topic of advisory committees is not without some good theoretical pieces. See especially Mort Grant, "The Technology of Advisory Committees," *Public Policy* (1960): 92–108, and Carl Kaysen, "Model-Makers and Decision-Makers: Economists and the Policy Process," *Public Interest* (Summer 1968) 80–95. They show the value of segmenting the advisory process by issue area for a greater understanding of the role of advisors in government. Important empirical studies are Thomas E. Cronin and Norman C. Thomas, "Federal Advisory Processes: Advice and Discontent," *Science* 171 (26 February 1971): 771–79; Michael Lipsky, "Social Scientists and the Riot Commission," *Annals* 394 (1971) 72–83; and Lyle Groeneveld, Norman Koller, and Nicholas C. Mullins, "The Advisors of the United States National Science Foundation," *Social Studies of Science* 5 (1975): 343–54. See also my "Science Policy in American State Government," *Minerva*, July 1971, 321–48 and "The Massachusetts Governor's Advisory Committee for Science and Technology: A Committee in Search of a Mission," August 1970 (distributed through the National Technical Information Service, PB-24-634 [Dec. 1973]).

ADVOCACY, NOT ADVICE

The Naval Research Advisory Committee did not enter fertile ground. Twice in its history, the Navy had convened committees of scientists approximately similar in purpose and stature to NRAC, and twice it had ignored them. In 1863 the Secretary of the Navy approved the formation of a Permanent Commission to advise the Navy on questions of science and appointed as members the three eminent scientists of the day who had urged its creation: Rear Adm. Charles Davis, Chief of the Bureau of Navigation; Alexander Dallas Bache, Superintendent of the Coast Survey; and Joseph Henry, Secretary of the Smithsonian Institution.[10] The Permanent Commission (which of course was not permanent) consumed the two years of its existence reviewing the many useless inventions submitted by citizens to aid in the Civil War. The Navy prevented the Commission from inquiring about the design and construction of ironclad vessels, the only significant naval innovation to occur during the conflict. These were topics that the Secretary and senior officers wished to reserve entirely for departmental personnel.[11]

The next conflict of great consequence to engage the Navy—the First World War—also led to the establishment of a science advisory committee, this one known as the Naval Consulting Board. Created in 1915 at the insistence of Thomas A. Edison, the board was chaired by Edison and consisted of two members from each of eleven national technical societies. Again the Navy burdened it with the task of screening the estimated forty thousand inventions that were submitted between 1915 and 1917. The only memorable contribution of the board was the suggestion that the Navy establish a central laboratory equivalent to those already found in industry for the exploratory development of new concepts. The Naval Experimental and Research Laboratory, today's Naval Research Laboratory, was commissioned in 1923, several years after the end of the war and after the quiet demise of the Naval Consulting Board.[12] A Washington, D.C. location, rather

[10] I. Bernard Cohen, "American Physicists at War: From the Revolution to the World Wars," *American Journal of Physics*, August 1945, 230.

[11] Nathan Reingold, "Science in the Civil War: The Permanent Commission of the Navy Department," *Isis* 49 (September 1958): 307–18.

[12] Rear Adm. Julius A. Furer, USN (Ret.), *Administrative History of the Navy Department in World War II* (Washington, D.C.: Department of the Navy, 1959), 753, and the committee's official history, Lloyd Scott, *The Naval Consulting Board of the United States* (Washington, D.C.: Government Printing Office, 1920). Also, O. W. Helm, "Genesis of the Naval Research Advisory Committee," *Research Reviews*, November 1957, 20–29. The British experience with science advice during the First World War is discussed in Roy M. MacLeod and E. Kay Andrews, "Science Advice in the War

than one in New Jersey, was selected for the laboratory in order to limit the Wizard of Menlo Park's influence on its operations.

It was appropriate that the man who had attempted to block the efforts of scientists volunteering to aid the development of naval weapons at the start of the Second World War, and who was himself relegated to sorting inventions during that war, would, at its end, be instrumental in establishing yet another science advisory committee for the Navy. Admiral Bowen, who in 1946 was in the process of converting his wartime ORI into the ONR, seized upon the suggestion, apparently independently offered by Adm. Luis de Florez and Vannevar Bush,[13] to assemble another committee of scientific consultants.[14] A provision for such a committee was duly inserted into the bill pending before Congress, calling for the creation of the ONR. Even before the bill passed, discussions were held with leading scientists to select possible candidates for membership.[15]

Bowen's purpose in seeking the establishment of a panel of scientific advisers was to bolster his claim within the Navy Department to the mission to develop nuclear power plants for naval vessels. The internal memoranda on the committee's membership and even the newspaper reports of the panel's establishment refer to it as the Navy's consulting board on nuclear power.[16] But by the time the bill became law, Bowen had lost the fight to make ONR the driving force in building the nuclear navy, as well as his opportunity to remain in uniform. Although there were several nuclear physicists on the new committee, it was not nuclear propulsion for naval vessels that they were to discuss.

The legacy Admiral Bowen left the Navy was an organization dispensing research contracts to academic scientists, most of whom were thoroughly confused by its seemingly disinterested generosity. So too were members of the committee (led by their chairman, Warren Weaver), who began to question at the committee's earliest sessions the intent and intensity of the effort. Why was the program so ill defined, he asked. Was the research being sponsored really relevant to the Navy? What will happen to the program if and when a national

---

at Sea, 1915–1917: The Board of Invention and Research," *Journal of Contemporary History* 6, no. 2 (1971): 3–40.

[13] Interview document.

[14] Helm, *Genesis*, 23.

[15] Ibid., and memorandum, to: Chief, Office of Research and Inventions; from: Head, Scientific Branch; subject: Possible Recommendations for Members of a Consulting Board to the Navy on Nuclear Power Development; 12 April 1946.

[16] Memorandum, to: Assistant Secretary of the Navy; from: Chief, Research and Inventions; subject: Possible Recommendations for Membership of a Consulting Board to the Navy; 29 April 1946.

science foundation is established? And why was ONR spending money so much faster than the Office of Scientific Research and Development which at least had a war to prepare for? "Does the activity suffer from a feeling on the part of operating personnel that, having asked for x million dollars, they had jolly well spend x million dollars? I ask this because I am, at the moment, dazed by the rapidity with which the money is being committed."[17]

The staff response was to assure the committee that ONR's program was well conceived. Its purpose, the staff argued, was to enhance long-range naval capabilities through advances in basic research. Nearly alone, ONR was carrying on work that the Office of Scientific Research and Development had initiated. The war had created a shortage of scientific manpower and of scientific research "which the ONR program [is] . . . alleviating."[18] The establishment of a national science foundation, it was noted, was not certain and could be postponed for years. If it were established, ONR would certainly pass on a considerable portion of the contracts to the foundation. In the meantime, though, it would be helpful if the committee not only provided advice in the setting of research priorities, but also joined the effort to "educate" the scientific community, Congress, and the Bureau of the Budget on "(a) the ultimate value to the Navy and Nation of fundamental scientific research, and (b) the interim NSF character of the work of this office."[19]

In fact, Admiral Lee (who had replaced Admiral Bowen as Chief of Naval Research) was soon quite explicit about the assistance desired of NRAC. In January 1947 the Bureau of the Budget proposed a major cut in ONR's program that possibly would not be made up through reallocations within the Navy's budget. The Admiral noted that while he was bound by an Executive Order not to testify before Congress against any presidential budget recommendation, the committee's members were not. Without the committee's support there might not be funds for the universities which could be very embarrassing, as the universities had just been told by ONR that the Navy was now in the business of financing academic research.[20]

When Admiral Lee reported two months later that the threat of a budget cut was serious, the committee made its choice. If a budget

[17] "Comments by Planning Divisions, ONR, on Questions Posed by Dr. Warren Weaver, Naval Research Advisory Committee, 12 November 1946," mimeograph, NRAC files.

[18] Ibid., 4.

[19] Ibid., 5.

[20] Transcript of the second meeting of the Naval Research Advisory Committee, Washington, D.C., 15 January 1947.

cut was imposed, NRAC favored trimming the university research program more severely than the Navy laboratories which ONR supervised.[21] The preference, however, was that the budget for research be increased, not reduced. The committee's decision was to recommend to Congress that the exploratory research share of the Navy's budget be raised from .5 percent to 2 percent, a figure it argued would be more in line with the practice of industry. If any reductions were to be made in the overall Navy budget, the committee recommended that research allocations be exempt.[22]

Despite this willingness to endorse an enhanced role for research in the Navy, some members continued to doubt the wisdom of utilizing military justifications for research support. Bronk, for example, thought "that much of the research sponsored by ONR should be financed through other channels [e.g., a civilian science foundation]. It would seem more reasonable if it were so. Some scientists would sleep with an easier conscience and many scientists would be better satisfied with their traditional environmental [sic] status if such were the case."[23] He hastened to add that ONR had treated scientists wonderfully, that there had been no restrictions on their research, and that he knew that there was Navy-relevant research that had to be done. Still, a federal science foundation seemed to Bronk the more appropriate mechanism for much of the work that ONR was sponsoring.

The committee, however, immediately began to close ranks behind ONR. Dr. Richard Dearborn, Texaco's Vice President for Development and a long-time friend of senior Navy Department officials, argued that the Navy needed a broad involvement with science and engineering and that the ONR "should continue to sponsor research and not be held down to specific Navy problems."[24] Similar comments were offered by other members. Thus, toward the end of the session, when Admiral Nimitz (the Chief of Naval Operations) entered the room, harmony prevailed. Assistant Secretary Brown, who had been present for all of the discussion, summarized the committee's view for Nimitz: "The Committee [is making] an eloquent plea for restitution of funds for ONR." Bronk joined in by noting for Nimitz, who had to leave immediately for another appointment, that not only was ONR sponsoring first-rate research, but it was also building a bank of good-will with eminent scientists which could have important future repercussions upon naval activities. Assistant Secretary Brown closed the session by

[21] Transcript of the third meeting of the Naval Research Advisory Committee, Washington, D.C., 26 March 1947.
[22] Ibid.
[23] Ibid.
[24] Ibid.

reminding the members that they, unlike him, were not bound by the precept to support the president's budget.[25]

The doubts kept reoccurring. At a meeting of NRAC in early 1948, Dearborn himself raised them, though he was careful to disclaim holding any of his own, noting instead that friends in industry had been inquiring about the Navy's priorities which were forcing reductions in development projects while leaving research untouched. "Is it not the function of Congress," he wondered, "to finally determine whether or not a science foundation bill shall be passed? May we not be subject to just criticism if we go on and do not act soon on the matter of which of the [research] projects are of real interest to the government and which are not."[26] Robert Oppenheimer, the Director of the Institute for Advanced Study and the scientific leader of the project to develop the atomic bomb, commented upon joining the committee in 1949, that there seemed to be an overemphasis on nuclear physics in ONR's research program with much of the work unlikely ever to have any relevance to naval applications. It was time, he thought, to let someone else worry about the support of basic research in that field.[27]

But NRAC nevertheless continued to endorse ONR's program and assist in the effort to increase its appropriation. Excerpts from its September 1949 meeting provide but one example of its willingness to be ONR's advocate:

Dr. Arthur Compton: I would be glad to move that we go on record . . . calling attention to the [fact that the—transcript garbled] level of research done by ONR seems inadequate to the present needs of the organization.
Would you take exception to that Admiral?

Admiral Solberg [Chief of Naval Research]: No, I was thinking along the same line. If we could have something positive in there, that you viewed with alarm the reduction from the 1950 budget by 10 percent in research, and that further reductions in years to come would be rather disastrous, that would be good. . . . You remember, Dr. Compton, two meetings ago I think it was, you said we should be spending twice as much.

Dr. Compton: And I would still support that.[28]

[25] Ibid.

[26] Transcript of the sixth meeting of the Naval Research Advisory Committee, Washington, D.C., 20 January 1948.

[27] Transcript of the ninth meeting of the Naval Research Advisory Committee, Washington, D.C., 9 June 1949.

[28] Transcript of the tenth meeting of the Naval Research Advisory Committee, Washington, D.C., 19 September 1949.

During the same session, the Admiral prepared NRAC for a meeting it would have with the Under Secretary of the Navy, Daniel Kimball. "I want to warn you in advance," he said, "that the Secretary knows relatively little about ONR. I have never been able to get in to talk with him about it, so I would like to give you a tip to educate him on the Office of Naval Research, its functions, its philosophy, and how you feel about it. It will make it easier for me later on."[29]

Not surprisingly, by the time NSF was created, NRAC had already committed itself to support ONR's continued support of basic research. Certainly there was to be a role for NSF, perhaps one concentrated on science education, statistics gathering, and the institutional support of science; but ONR, the committee asserted, should have a totally independent and undiminished opportunity to sponsor basic research. The resolution recording NRAC's position was moved by Robert Oppenheimer and seconded by Admiral Lewis Strauss, who would soon become Oppenheimer's bitter antagonist in debates in the General Advisory Committee of the Atomic Energy Commission over national security policy. In this instance, however, they were united in the common cause to guarantee pluralism (and ONR's continued existence) in the support of science.[30] Over the next two decades, although its membership constantly changed, NRAC remained the faithful advocate—writing letters to senior officials, testifying before Congress, and using personal contacts in and out of government—all in the effort to keep the budget for ONR's basic research program expanding. NRAC never failed a request from ONR to assist in budget battles and intervened on its own initiative whenever ONR's interest in basic research appeared to falter.[31]

DISCOVERING THE CONSTRAINTS

Predictably, eminent scientists who convene quarterly to renew their faith in science soon look for other tasks. This would appear to be especially true if they are brought together by the military at a time when their nation is at war. The frustrations burst forth at the NRAC meeting of 25 June 1951, a year after the Korean War began; the committee wanted something else to do besides endorsing ONR's budget aspirations. As Robert Oppenheimer expressed it with his character-

[29] Ibid.

[30] Transcript of the eleventh meeting of the Naval Research Advisory Committee, Washington, D.C., 30 January 1950.

[31] Examples are found in transcripts for the twelfth, nineteenth, twentieth, thirty-first, thirty-third, thirty-fifth, thirty-sixth, thirty-seventh, and fortieth meetings of the Naval Research Advisory Committee.

istic bluntness: "What in the hell are we here for? . . . We have said every time that it was a very successful program, but I think it has occurred to us that there may be individual members of the committee who have some special competence, or special interest [that] might be better [utilized] in some aspect of the program. If it could be sent to us in advance maybe of a meeting then possibly a meeting of the Admiral or his assistants who are concerned with it might be willing to allow us to discuss it then before the whole committee and some new ideas or some contribution might develop."[32]

The Office of Naval Research reacted by providing NRAC with frequent briefings on naval weapons research and occasional field trips to major naval installations. For example, at the next NRAC session (the meeting of October 1951) representatives from the Bureau of Aeronautics gave a presentation on aircraft carrier launch and recovery techniques.[33] A subsequent meeting included demonstration movies of the F9F and various missiles. Although ONR gave NRAC no new issues to explore, the committee was content, for a while at least, to be kept informed of progress in the development of naval weapons.[34]

By late 1955, however, the committee had grown impatient with ONR's diversionary tactics. NRAC's new chairman (the committee elected its own at that time, but now that post is filled by the nomination of the Assistant Secretary of the Navy for Research, Engineering and Science), Dr. Julius Stratton of M.I.T., made it clear that he expected the committee to do more than it had in the past. The Air Force, he said, was making better use of its Science Advisory Board (SAB) and it was time NRAC itself should go to work.[35] One member of the committee, Dr. Mervyn Kelly (who was also on the Air Force's advisory panel), sought to alleviate any fears that the Navy officials present might have had that scientists were not sophisticated enough to deal with important policy issues. "I assure you," he said, "we are very careful in the political sense, and in the technical sense, to see that the recommendations [of the SAB] are sound and not embarrassing to the Air Force. Our judgment and [General] Putt's are combined

[32] Transcript of the fifteenth meeting of the Naval Research Advisory Committee, Washington, D.C., 25 June 1951.

[33] Transcript of the sixteenth meeting of the Naval Research Advisory Committee, Washington, D.C., 22 October 1951.

[34] NRAC's chairman noted after F9F movie, "it certainly is more than pleasant to see the results of the research proceeding and progressing into areas where they are going over into the applied phase and meeting the ultimate purpose." Transcript of the twenty-second meeting of the Naval Research Advisory Committee, Washington, D.C., 18 March 1954.

[35] Transcript of the twenty-sixth meeting of the Naval Research Advisory Committee (held at the U.S. Naval Mine Defense Laboratory), 9 November 1955.

on that, and if he disagrees with a thing to be set up, we hold our heads a long time before we send it up [to the Chief of Staff and the Secretary]."[36] It is possible to wonder how this comment was received by the officials, as many of the policy issues which absorbed their time involved disputes with the other armed services, not the least among them the Air Force, which Dr. Kelly and other NRAC members advised so discreetly.

The Naval Research Advisory Committee had already become somewhat busier, but not with the big issues. The committee had helped select a new research director for NRL. Two of its members had accompanied the Assistant Secretary of the Navy for Air on a European study tour. And the chairman did summarize for ONR the committee's view that the Navy's problem in mine warfare was not research investments, but rather the lack of a consensus on the role of mines in warfare.[37]

Apparently though, NRAC's dissatisfaction did reach the top. Its next meeting (the first in 1956) was attended by Assistant Secretary Smith, the CNO, Admiral Burke, two vice admirals, ten rear admirals, and fifteen captains, and featured an invitation for NRAC to look into antisubmarine warfare and air defense at sea—problems that were then and still are at the center of the Navy's agenda.[38] Shortly thereafter, the NRAC received a letter from the Secretary of the Navy reconfirming the committee's powers and notifying it of the availability of staff support in the form of an executive director. The briefings that the Navy offered immediately became more substantive.[39]

With its new activity, NRAC did not lose interest in the vitality of basic research. Periodically, it would issue statements about the need to increase the Navy's support for research.[40] However, the commit-

[36] Ibid.

[37] Letter from Chairman Stratton to Capt. Meyers, dated 16 November 1955.

[38] Transcript of the twenty-seventh meeting of the Naval Research Advisory Committee, Washington, D.C., 17 January 1956. The turning point came when Bronk wrote Admiral Burke in early December 1955 proposing the establishment of a CNO advisory committee. Admiral Burke's reply, later that month, noted that NRAC already held responsibility to advise him as well as the CNR. The leadership of NRAC interpreted this to mean that they held the license to act as the CNO's adviser and that the next move was up to NRAC. Minutes of an informal meeting of the Committee on Undersea Warfare of the National Research Council held at Lyman Laboratory, Harvard University, 5 January 1956.

[39] Transcript of the twenty-eighth meeting of the Naval Research Advisory Committee, Washington, D.C., 4 May 1956 and of the twenty-ninth meeting of the Naval Research Advisory Committee, 14 June 1956.

[40] Transcript of the thirty-fourth meeting of the Naval Research Advisory Committee, Washington, D.C., 22 April 1957, and transcript of the thirty-fifth meeting of the Naval Research Advisory Committee, Washington, D.C., 28 October 1957.

tee's patience for dealing with ONR's problems was obviously beginning to fade. When Admiral Bennett, the Chief of Naval Research, asked for assistance in choosing between more money for solid-state physics or investments in two or three additional particle accelerators, he was told that it was a foolish question. Several members suggested that instead he should have asked what was the supply of good people in these fields so that ONR could get them the resources that they would need to do first-rate research. The Admiral's protest that it might be difficult to get the necessary funds from Congress was answered by comments to the effect that it may be more the temerity of the agencies in requesting support rather than the frugality of Congress in appropriating funds that kept the budget for basic research limited.[41]

The Naval Research Advisory Committee had another concern; its first big policy issue. In October 1958 the committee was asked by the Secretary of the Navy on behalf of Admiral Burke to prepare a study of the optimal size aircraft carrier for the Navy. NRAC was reluctant to take on the task because it feared, rightly so, that the committee was being asked to endorse large carriers, the controversial centerpiece of U.S. naval strategy.[42] Other naval officials pressed the issue by providing briefings on the air and submarine threats to carriers.[43] Impressed by the apparent vulnerability of aircraft carriers, the committee began to side with those opposed to a large carrier strategy.[44] However, an attempt by NRAC to take a formal position on the issue was scuttled when it became clear that senior admirals believed that such an action by the committee could jeopardize the Navy's appropriation request, which included funds for carrier construction and which was pending in Congress.[45] Although two or three drafts of a NRAC policy statement on carriers were prepared, they were not submitted to the Secretary "on advice from across the river."[46] The Navy already had an official

[41] Ibid., thirty-fifth meeting. Given the fact that Sputnik had just been launched when this discussion occurred, the NRAC members may have been correct in their assessment of the budgetary politics.

[42] Transcript of the thirty-ninth meeting of the Naval Research Advisory Committee, Washington, D.C., 23 October 1958.

[43] Transcript of the forty-third meeting of the Naval Research Advisory Committee, Washington, D.C., 22 October 1959.

[44] Letter to the Secretary of the Navy from the Chairman of the NRAC, 17 February 1960.

[45] Transcript of the forty-fifth meeting of the Naval Research Advisory Committee, Washington, D.C., 2–3 May 1960.

[46] Transcript of the fifty-fifth meeting of the Naval Research Advisory Committee, Washington, D.C., 8 November 1962.

position on carriers; it was in favor of them and it liked them large. NRAC was soon back to reviewing ONR's basic research plans.

THE SUMMER AND WINTER OF ADVICE

The Naval Research Advisory Committee, to be sure, was not the Navy's only source of scientific advice. The Navy occasionally initiates special scientific panels, often referred to as summer studies because they hold continuous sessions during the academic summer recess, to investigate weapon strategies and opportunities. It also has the continuing assistance of the Naval Studies Board of the National Research Council, the operating arm of the NAS.[47] And it has established the Center for Naval Analyses as a contract study unit along the lines of the Air Force's well-known RAND Corporation. Most of the advisory mechanisms are administered, though not necessarily controlled, by ONR.

Scientists tend to revere summer studies as effective devices for influencing national security policy.[48] Some summer studies certainly can be said to have had an important impact upon policy. For example, the DEW line early warning system grew out of the recommendations of a 1952 study, and the design for the Polaris missile gained impetus from the Project Nobska study of 1956. But other studies, though they may have dealt with significant policy issues, have left no mark upon policy. The strength of the studies often lay less in the originality of the ideas they presented than in the channel to the top of the military establishment that they provided for ideas long stifled in the lower levels of the military. The reiteration and endorsement by a panel composed largely of prominent scientists legitimized ideas and forced their review, if not adoption, by senior officials.[49]

As interesting as the achievements of summer studies, are their origins. When the Cold War became a reality in the late 1940s, certain scientists, Vannevar Bush among them, pressed for a greater role in weapon development decisions than that which they had been given. Bush, in particular, thought it would be appropriate to reorganize the Joint Chiefs of Staffs along the lines of their Canadian equivalents to

[47] Mark Bello, "Armed Forces Receive Advice on National Security Needs," *National Research Council News Report* 37, no. 7 (July 1987): 6–9.

[48] For example, Jerrold Zacharias, "Scientist as Advisor," paper presented to the Science Policy Seminar, Harvard University, Cambridge, Massachusetts, 29 March 1961, and J. R. Marvin and F. J. Weyl, "The Summer Study," *Naval Research Reviews*, August 1966, 1–12.

[49] Interview documents. Transcript of the forty-second meeting of the Naval Research Advisory Committee, Washington, D.C., 27 July 1959.

include a civilian scientist among their members.[50] The military chiefs, never enamored with scientists, were appalled by the suggestion. Hardly more enthusiastic was their reception to a proposal Bush made after the start of the Korean War entitled "A Few Quick" that called for the re-creation of an OSRD-type organization to promote weapon innovations independent of the military.[51] The summer study, tentatively tried for the exploration of a few technical questions, was seized upon as the mechanism to contain these pressures. Scientists could have a role, but it would be one organized and controlled by the military.[52]

Co-optation was an efficacious strategy. Many scientists believed that thirty to forty of their brightest, assembled for a summer, could solve the most difficult policy problems. Often, exciting ideas did result when fresh perspectives were applied to old issues by the scientists. The military knew, however, that brilliance is hardly ever the match for bureaucracy. Whatever the recommendations of a study group might be, it would be the military that would control their implementation. A part-time rival is always less of a threat than a full-time one.

Summer studies gradually faded in importance. Scientific endorsements grew trite and seemed too obviously manipulated by the military. The antagonism generated by the Vietnam War made it difficult to recruit participants. And the growth of scientific expertise within the government and specially designated contract organizations appeared to diminish both the desire and the appropriateness of seeking guidance from outsiders, even those with years of experience in weapon technologies.[53]

Studies prepared by contract organizations took the place of those prepared by scientific panels in the policy debates. The first contract organization formed, the RAND Corporation, was a creation of the Air Force. RAND's hallmark was the application of quantitative techniques such as operations research and cost/benefit analysis to defense ques-

---

[50] Interview document.

[51] Memorandum for Adm. C. M. Bolster; subject: notes on V. Bush's extensive memorandum, "A Few Quick," dated 5 November 1951; 7 January 1952. ONR staff thought that if any organization was to be given the "Few Quick" mission to develop innovative technology for the military, it should be ONR. Transcript of the seventeenth meeting of the Naval Research Advisory Committee, Washington, D.C., 19 March 1952. Note also Furer's comment, "The establishment of the Office of Naval Research with bureau status and funds of its own is another factor reducing the likelihood that an OSRD type of mobilization will be repeated," 806.

[52] Transcript of the fourteenth meeting of the Naval Research Advisory Committee, Washington, D.C., 19 March 1951.

[53] Interview documents.

tions.[54] Many people in and out of government were impressed with this approach to defense problems, especially as it appeared to offer "scientific" solutions to difficult political choices. The Navy, despite the fact that it had helped develop at least some of these techniques during the Second World War,[55] felt at a disadvantage in policy debates because it lacked a close identification with an organization like RAND that could argue the Navy's case in analytical terms while maintaining broad creditability within the academic community. As RAND's influence grew within the Department of Defense during the late 1950s and early 1960s, so did the desire within the Navy to acquire its own RAND equivalent.

Two obstacles stood in the way of a Navy RAND. First, because of its own internal fragmentation, the Navy soon found itself with a half-dozen analysis units producing or commissioning competing and, at times, conflicting studies. There was a naval analysis group in Cambridge, Massachusetts; another at the War College in Newport, Rhode Island; a couple in Washington, D.C.; and a few others scattered around at the naval laboratories. Each worked for a different naval staff unit. Second, the Navy was reluctant to provide any of the units with the independence and visibility necessary to gain a reputation for effective work. As the director of one of these units put it at a NRAC meeting in 1962: "The Navy is sorry it doesn't get the kind of publicity that RAND has brought to the Air Force, but let you try to get some of this publicity and see how far you get. The Navy is schizophrenic about public relations . . . it goes for it in principle but not in practice, or rather, there are grave limitations to how far it will go. Part of it is the general idea 'Don't give ammunition to your enemies.' "[56]

The advent of the McNamara regime in the Department of Defense forced the Navy to consolidate its analysis units into a single structure, if only to defend itself more effectively against a Secretary of Defense who was so totally enamored with studies. In 1962 the Navy created the Center for Naval Analyses (CNA), bringing together the Operations Evaluation Group (OEG), the direct descendant of the an-

[54] Bruce L. R. Smith, *The RAND Corporation: Case Study of a Nonprofit Advisory Corporation* (Cambridge, Mass.: Harvard University Press, 1966). See also Philip Green, "Science, Government and the Case of RAND," *World Politics* 20, no. 2 (January 1968): 301–26. Note also I. B. Holley, Jr., "The Evolution of Operations Research and Its Impact on the Military Establishment: The Air Force Experience," in *Science, Technology and Warfare*, the proceedings of the Third Military History Symposium, USAF Academy, 1969, 89–109.

[55] L. Edgar Prina, "The Navy First Used Think Tanks During World War II," *Armed Forces Journal* 106 (28 September 1968): 18–23.

[56] Transcript of the fifty-second meeting of the Naval Research Advisory Committee, Washington, D.C., 9 February 1962.

alytical efforts begun within the Navy during the Second World War, the Naval Weapons Analysis Group (NAVWAG), a mid- to long-range planning unit established to assist OPNAV; and the Institute for Naval Studies (INS), itself a consolidation of several analytical units and the product of an earlier effort to build a Navy RAND.[57] The Institute for Defense Analyses (which had held the contract for INS), M.I.T. (which had held the contract for OEG and NAVWAG), and the Smithsonian Institution (which conveniently had a board of leading politicians), were all approached to be the management contractor for CNA, but declined for various reasons. The Franklin Institute finally agreed to take on the task, only to be replaced a few years later by the University of Rochester and then by the Hudson Institute. ONR was the formal contracting agent for the Navy, although program direction was held by OPNAV. Activation of CNA was slowed by conflict among the different organizational cultures of its component units and by their scattered locations. Eventually, CNA selected a Virginia suburb of Washington as its headquarters and prime location.

Although CNA has done some important studies, its reputation for objectivity has never matched that of RAND, either within or outside the government. From the beginning, the Navy chose to staff CNA heavily with naval officers on the rationale that their involvement with analytical studies would improve the relevance of the work as well as help create a cadre of analytically skilled officers.[58] Some observers, however, believed the officers to be the Navy's version of political commissars, assigned to CNA to assure that its studies conformed to OPNAV established positions. OPNAV, it was felt, has never been ready to risk the free analysis of its programs. Rather, it wanted only an organizational champion to promote the Navy's cause in policy debates increasingly conducted in the language of analysis.[59] That the champion created was intensely loyal was not unexpected. And just in case it was not, requirements were imposed that all of its studies (and those of any other analytical group working for the Navy) be reviewed and approved by OPNAV before release to the Department of Defense.[60] A

[57] Rear Adm. Edwin B. Hooper, USN, "Early History of the Institute of Naval Studies," January 1964 (internal unpublished manuscript); memorandum from ONR/460S to CNO OP-07T; subject: Historical Background on Management and Coordination of Navy Studies and Analyses. The creation of CNA is fully described in Keith R. Tidman, *The Operations Evaluation Group: A History of Naval Operations Analysis* (Annapolis: Naval Institute Press, 1984), chap. 4.

[58] Tidman, *The Operations Evaluation Group*; Interview document.

[59] Transcript of the fiftieth meeting of the Naval Research Advisory Committee, London, England, 13 September 1961.

[60] See CNO memo ser 022P90 of 7 March 1963, Department of Navy Studies to be submitted to SECDEF; OPNAVINST 5000.30 of 13 January 1964, Coordination and Super-

special OPNAV office, OP-96, the Systems Analysis Division (now re-
numbered and renamed OP-81, the Program Resource Appraisal Division) was established in 1966 to oversee the studies and conduct
some of its own.[61]

CONCENTRIC CIRCLES

No one should be surprised that OPNAV would be wary of outside advice, even that which it solicits. OPNAV is the political core of the Navy.
It helps organize the coalition of bureaucratic interests that defines
the Navy's interests. It defends the Navy from attack, civilian or military, foreign or domestic. And it seeks to preserve the Navy as an independent, viable organization that will be able to provide interesting
and predictable career opportunities for naval officers.

Naval officers practice the profession of conducting warfare at sea.
Similar to other professionals, they have a special knowledge, enduring norms of behavior, a desire for autonomy, and collective as well as
individual aspirations. But unlike most other professionals, they have
only one organization in which to fulfill their aspirations. The fate of
that organization is the fate of the profession.

Trust is extended to others to the degree to which they share the
common fate. The careers of scientists, including those with years of
experience working on naval problems, are but rarely tied totally to
the Navy's prospects. Some, thinking those prospects to be irrelevant
or of secondary importance, advocate defense policies that are potentially threatening to the Navy's interests, as was the case with the proposal for the re-creation of OSRD offered during the Korean War. Others may wish to define the Navy's interests on their own, without ever

---

vision of Studies Conducted Within and for Offices of OPNAV Relating to Navy Pro-
grams; CNO letter ser. 199P91 of 15 May 1964; subject: Coordination of Navy Department Studies; CNO letter ser. 22191 of 12 June 1964 to SECNAV; SECNAVINST 5000.23 of
29 June 1964, Coordination of Navy Department Studies; OPNAVINST 5000.30A of 17
October 1964, Coordination of Navy Department Studies; OPNAVINST 5000.29B of 23
February 1965, Functions and Organization of the Center for Naval Analyses and its
Relationship with the Department of the Navy; OPNAVINST 5000.32 of 21 April 1967,
Reporting In-House and Contract Operations Analysis Studies to the SECDEF, Procedures for; SECNAVINST 5000.23A of 26 July 1969, Department of the Navy Studies and
Analyses, Policies and Responsibilities for; OPNAVINST 5000.37 of 26 December 1969.
The contract with the University of Rochester had several clauses to ensure the inde-
pendence of the Center for Naval Analyses but doubts persisted regarding their effectiveness in preventing the Navy from controlling the work and conclusions of the organization. James R. Craig, "Captive 'Think Tank'—DOD $265-Million Hobby Horse?"
Armed Forces Journal International 113 (December 1975): 20–22.

[61] Comdr. Bruce R. Linder, USN, "Ops Analysis: Just 'Quantitative Common Sense,' "
Proceedings of the U.S. Naval Institute (August 1988): 98–101.

being subject to the organizational risks of their actions, as perhaps NRAC members were tempted to do when they considered taking a public position on the question of a large carrier strategy.

Contract advisors are only slightly more trustworthy. Their dependence upon the Navy is largely financial and likely to be only temporary. There is always the danger that they may adopt a neutral stance toward the Navy, providing political opponents with the most desirable of weapons, a Navy-sponsored study that does not support Navy policy.

Even civilian service scientists working in the Department of the Navy can be viewed with suspicion. Their career opportunities are broader than those of most naval officers. Their organizational loyalty is to the department and its various technical subunits such as the naval laboratories and weapon research centers, not to the U.S. Navy, which is (along with the U.S. Marine Corps) a military subunit of the department and composed exclusively of officers and enlisted personnel.

There are several distinctions that exist within the officer corps. The Navy needs specialists, but segregates them into service corps, e.g., Supply Corps, Medical Corps, Dental Corps, Civil Engineer Corps, and Chaplain Corps, undoubtedly because of their strong professional identifications. The Navy also needs line officers who are professionally qualified in its most important technologies. Those officers who specialize in these fields and who wish only shore assignments transfer into or are appointed to what is called the Restricted Line (e.g., Engineering Duty, Aeronautical Engineering Duty) and follow separate and well-defined careers that exclude them from combat command responsibilities and thus from the most senior ranks in the Navy (except in unusual cases, such as that of Admiral Rickover). Officers in the Unrestricted Line fly the planes, drive the ships, fight the wars, and have the least civilian career options. They are the ones who stand warily at the Navy's center and worry most about what others might suggest to change the Navy.

THE ABSENCE OF INDEPENDENCE AND RESPONSIBILITY

For Vannevar Bush the central issue in the relationship between science and the military was the need for civilian scientists to have an independent role in the formulation of defense policy. He believed that free men who wished to remain free would have to do research in the implements of war. He also believed that for such research to be effective, the active participation of at least a portion of the academic community would be required. And he was concerned that the decisions determining the type and function of weapons to be developed would become the monopoly of the military after the Second World War (as they were before that war).

At the end of the Second World War, Bush tried in a number of ways to ensure that scientists would have an independent voice in future defense decisions. First, he endorsed the proposal to establish within the National Academy of Sciences a Research Board for National Security that would manage military research. Next, Bush included in his own proposal for what was to become the NSF, a provision giving the civilian-directed foundation responsibility for "long-range research on military matters." Later, he helped organize a civilian-led Joint Research and Development Board in the Pentagon to coordinate postwar military research. Finally, when the Korean War broke out, he sought the reestablishment of OSRD in the "A Few Quick" proposal. All of these initiatives went awry largely due to counter maneuvers by the military.

But the most effective block to the creation of an independent base for scientists to assess the opportunities and need for weapons was the establishment of science advisory committees such as NRAC and the Air Force's Science Advisory Board. The existence of these committees symbolized to Congress and senior government officials the military's concern for technological innovation. Their members, mainly university presidents and leading scientists, willingly accepted the task of nurturing the military's investments in basic research, investments they were reluctant to jeopardize. Thus, they joined the military in opposing Bush's schemes and accepted limitations on their own advisory roles when their sponsoring agency's budgetary interests were at risk, as occurred in the case of NRAC's attempt to comment on the Navy's large carrier strategy.

Lacking the equivalent of an OSRD, the military has controlled the research support and the access to information of academic scientists interested in defense affairs since the Second World War. The management of the advice that the services received was a simple and organizationally irresistable extension of this control. Thus it is not surprising that when the military's relationship with science was examined anew in the late 1960s, much of the criticism emanating from the universities was ill informed. The result was a desire on the part of both the military and academic science to seek complete isolation from one another, a desire that if fulfilled, would not have served the nation or its defenses well.

CHAPTER SEVEN

Conclusion

THE Office of Naval Research helped American universities enhance their position of world leadership in science in the years immediately following the Second World War. The agency functioned as a surrogate national science foundation precisely during the period when basic research and graduate education in science and engineering needed federal nurturing the most. The war had totally disrupted European science, providing the opportunity for greater American assertion in basic research and advanced training. Hundreds of thousands of veterans were entering the universities eager to make up for lost time. There were dozens of important ideas in science and technology waiting to be developed. And ONR had the funds available and the administrative imagination required to allow American science to surge forward.

Two explanations are commonly offered for ONR's special role in postwar science, but neither is correct. Scientist statesmen looking for other agencies to step forward usually describe the role as both intended and exemplar. The Navy, in this view, recognized the great contributions science made to the war effort and established ONR to invest in the nation's future. Science, they argue, is a bountiful reservoir that the wisest agencies, not always able to anticipate future needs, help to keep filled.[1] Revisionist historians take a somewhat different view. The Navy, hoping to capture the future, set out to capture science—the source of future technologies—by establishing ONR to manage science's postwar renewal. The militarization of science was the result, they believe.[2]

But as described above, ONR's role was nothing more than a bureaucratic accident. Admiral Bowen had pressed for the creation of the office so that he might build the nuclear navy. Scientists, especially

---

[1] Harvey Brooks, "Basic Science and Agency Mission," in F. Joachim Weyl, ed., *Research in the Service of National Purpose: Proceedings of the Office of Naval Research Vicennial Convocation* (Washington, D.C.: Department of the Navy, 1966), 33–47.

[2] See, for example, Paul K. Hoch, "The Crystallization of a Strategic Alliance: American Physics and the Military in the 1940s" (unpublished manuscript). Sometimes government scientists and university advisers say the same thing. Note *Report of the Working Group on Basic Research in the Department of Defense*, Executive Office of the President, Office of Science and Technology Policy, 22 June 1978, V-3.

those returning to the universities from the atomic bomb project, were to be cultivated in order to gain access to the nation's nuclear knowledge, which was then being guarded by uncooperative Army officers. However, before Admiral Bowen could implement his plan, he was outmaneuvered for jurisdiction over nuclear propulsion by others in the Navy who made an accommodation with the Army in the broad range of defense policy and nuclear weapons issues. The Admiral quickly retired, leaving behind a law authorizing ONR, a budget, and a bewildered, but resourceful staff who found useful work supporting academic science in a most enlightened, if initially purposeless, manner. The rest of the Navy, busy demobilizing the fleet and arranging relationships in the newly-formed defense department, failed to notice what was happening to ONR.

Vannevar Bush, James Conant, and the other leaders of the wartime science effort were aware of ONR's activities, but paid little attention to them because they were absorbed in developing what they thought would be the grand design for postwar science. Of particular concern for Bush was that there be an independent role for scientists in the formulation of defense policies along the lines of what he thought he had achieved during the war. The military's unwavering resistance and many legal complications prevented this from happening, surely to the benefit of science.[3] Gaining a direct involvement in peacetime defense planning would have politicized science beyond recognition, requiring both an officially sanctioned hierarchy to manage internal conflict that would have been unacceptable to many scientists, and constant bureaucratic combat with the armed services over missions and strategies that would have been unwinnable.

What scientists were handed, thanks to ONR, was support nearly without interference. Money and equipment, not previously available except during war, found their way to the university laboratories. All that was asked was that the work be considered of high quality when reported in the literature. University administrators were persuaded to accept the funds by promises of full reimbursement for overhead expenses, support for graduate assistants, and other concessions that assuaged their fears of federal intervention. ONR's ability to clothe its allocations in the protective garb of national security justifications helped it avoid the political pitfalls of church vs. state, public vs. private, and geographic equity issues that had long been part of discussions of federal aid for educational institutions.

[3] For an excellent analysis of the issues that were central to the debate over the structure of postwar science policy, see Don K. Price, *Government and Science* (New York: New York University Press, 1954).

Senior naval officers did discover the office when the Navy experienced budget problems in the late 1940s. They did not believe that the Navy, through ONR, should be the generous and understanding patron of science that it had become, especially if it meant less money for weapon acquisitions. First one admiral and then another was sent to close down the operation. Only the flood of defense dollars that accompanied the start of the Korean War prevented ONR's early demise. Eventually, they and other defense officials did succeed in limiting ONR's budget for academic science, making the office focus much more than it had on naval technology needs and thus complement weapon acquisition plans.

Scientists did not protest the militarization of ONR. A growing disenchantment in the elite universities with defense policies, provoked largely by the Vietnam War, but also by nuclear weapon strategies as well, made a distancing from the military acceptable. So did the expansion of NIH and NSF budgets and the enactment of legislation providing for institutional subsidies for the universities and direct assistance for college students.[4]

In order to gain the research allocations, however, scientists have had to claim that their work directly serves current national needs. Two decades ago, science was concentrating its efforts on the space program and the cancer crusade. Later, the emphasis was on Project Independence, the attempt to reduce oil imports. Today science is focused on the AIDS epidemic and the struggle to make America economically competitive. When one national crisis fades, another must be found and promoted as solvable by science to ensure the flow of funds. American science may be relatively affluent, but this affluence is maintained at the price of being involved in an endless search for objectives.[5] Science is not militarized as some claim, but rather it is constantly being mobilized to serve ever changing national purposes.

National security rationales are no longer very important in the support of basic research. Today ONR ranks only sixth in the list of federal agencies financing academic science and, even after the Reagan defense buildup, it has barely returned to the budget levels in real terms that it attained in the late 1960s. The entire defense department provides less than a billion dollars for academic science, nowhere near the amounts generated by NIH and NSF. But without the protection of

[4] Advisory Commission on Intergovernmental Relations, *The Federal Role in the Federal System: The Dynamics of Growth*, vol. 6 of *The Evolution of a Problematic Partnership: The Feds and Higher Ed* (Washington, D.C., May 1981), A-82.

[5] John Maddox, "American Science: Endless Search for Objectives," *Daedalus* 101, no. 4 (fall 1972): 129–40. See also W. Henry Lambright, *Governing Science and Technology* (New York: Oxford University Press, 1976).

national security rationales, science is vulnerable to political pressures in ways that undermine its integrity and productivity. When vital defense interests are not at stake, politicians wonder why their districts are not benefiting from the federal research largess much more than when they are. Less favored institutions and disciplines find the urge to employ pork barrel tactics impossible to resist. The network of elites that binds together the scientific community and provides its priorities cannot contain the desire for equity and opportunity that is so much a part of the political process. The Navy and the other armed services may not regret their reduced role in basic research, but science no doubt will.

A SENATOR'S PROPHECY

A 1947 debate between Senator Leverett Saltonstall, Republican of Massachusetts, and two Democrats, Senators Richard Russell of Georgia and William Fulbright of Arkansas, over an amendment to the then pending NSF bill tells us much about our political system and the consequences that flowed from having ONR rather than a civilian agency allocating the initial postwar funds for academic research. The amendment would have required that a geographic distributional formula be used in the allocation of the foundation's grants. Senator Saltonstall argued against the amendment, citing the vital role M.I.T. and Harvard played in various weapon research projects. Senator Fulbright reminded the senator from Massachusetts that the issue before them involved "pure science research," not applied research, and thus more out-of-the-way places, as he put it, might also be able to contribute.

Senator Russell offered a prophecy: "I unhesitatingly put in the Record the prediction, here and now, that within six years after the pending bill shall have been enacted into law, in the absence of an amendment of this nature, the two institutions referred to will be receiving more funds than all the educational institutions in at least the 12 States west of the Mississippi River and south of the Potomac."[6] The bill passed without the amendment, only to be vetoed by President Truman on other grounds. It was not until 1950 that the foundation was established and not until the late 1950s that it received nontrivial appropriations. Adjusting the dates, Senator Russell correctly foretold the future. The gentleman from Georgia, however, also served on the Senate Appropriations Committee where he routinely approved funds for ONR which initially took the place of NSF and which

[6] *Congressional Record-Senate* 43, part 4, 16 May 1947, 5425.

gave much of its largess, safely protected by the presumption that weapons of some sort were being built, to m.i.t., Harvard, and a few other universities, none of which was located in the states bounded by the Mississippi and the Potomac.

Of course, onr was not immune to political pressures. It did manage to support an extra contract or two in Texas and perhaps even a few in Georgia as well.[7] But more often the politics to which it was sensitive was that of the basic research community, particularly as found in the elite universities.[8] It was always ready to support leading scientists in fields of interest. Scientists were given freedom in selecting topics to investigate and encouraged to include graduate students in their work. onr stood ready to assist in the acquisition of research equipment. It was the enlightened public patron American scientists had always sought, but had never previously located.

The research relationship that onr pioneered grew into the favored channel for infusing the universities with much greater amounts of aid as a college education became the societal norm for middle-class status.[9] National security rationales, again pioneered by onr, protected the assistance from the political scrutiny of the type desired by Senators Russell and Fulbright. Eventually, the demands for equity in the distribution of federal funds became so great that other mechanisms such as subsidized student loans and institutional grants supplemented the research contract as conduits for providing the universities with financial assistance.[10] At the same time, largely because of the Vietnam War, national security rationales lost their aura, and the research contracts upon which the elite universities were still heavily dependent became subject to the vagaries of democratic politics.

[7] Interview Document. In a memo to Captain Conrad dated 27 March 1946, Admiral Bowen notes that Representative Albert Thomas of Texas "is very interested in having educational institutions in his district closely hooked up with this Office. He even suggested having an onr representative in that area. This, however, would seem to me to be going a little bit too far." The Admiral did go on to instruct Captain Conrad to write to Thomas about his recent trip to Rice to let the congressman know that onr was paying attention to the district.

[8] See Daniel S. Greenberg, The Politics of Pure Science (New York: New American Library, 1967), and Earl J. McGrath, The Liberal Arts College and the Emergent Caste System (New York: Teachers College, Columbia University, 1966).

[9] Daniel S. Greenberg, "The New Politics of Science," Technology Review 69 (April 1967): 49–56, and Harold Orlans, ed., Science Policy and the University (Washington, D.C.: Brookings Institution, 1968).

[10] Harvey M. Sapolsky, "Science Policy," in F. Greenstein and N. Polsby, eds., Handbook of Political Science 6 (Reading, Mass.: Addison-Wesley, 1975) 79–110; Don K. Price, "Science and the Great Society," in B. H. Gross, ed., A Great Society (New York: Basic Books, 1968), 227–47.

Ridding the Campuses of Uniforms

The Office began a search for a naval mission one war before Vietnam. Although it had survived the trauma of Admiral Badger's inquiry, the agency could never thereafter forget that the Navy's interest in science was entirely pragmatic. More and more, ONR came to use the Navy's operational requirements as the guide for supporting projects.

First, the Department of Defense, then the President, and finally Congress reiterated the point that federal research allocations had to be linked closely with agency missions. ONR's budget stagnated in this drive for relevance, in large part because the office clung to the belief that it was necessary to maintain investment in basic research and to nurture graduate education. New weapons were the desired objective; and there was not enough evidence that ONR was producing new weapons at a sufficient rate to justify its budget.

It was ironic that during the campus turmoil of the late 1960s the defense research agencies, including ONR, were singled out for attack as supporting research that was too warlike. Only a few of the many scientists ONR had supported were moved to defend the agency to its radical critics even though ONR was still seeking to persuade an increasingly hostile officer corps of the benefits of unfettered research and flexible contractual relations.[11] Instead, many others, now safely linked to another government research sponsor, sought to cleanse the universities of the supposedly corrupting influence of the defense agencies. Like a number of senior officials in the military establishment that they opposed, these scientists wished to have their campuses free of uniforms.

One wonders who was acting more irresponsibly, the scientists who broke their ties to the military or the officials who were happy to see them do so? The assumption has to be that the society that seeks to be free will always seek to defend itself even though some members of a privileged class within it may convince themselves that armaments are the cause of conflicts and that defense dollars are tainted. A group of scientists denying the military their services is likely to be much less of a problem for such a society than having a military alienated from, and unknown by, a large segment of the university community. It would seem to indicate unusual arrogance to believe otherwise.

[11] For a statement by a scientist who did step forward, see James Case, "The University and Defense Research," *Naval Research Reviews* 23, no. 2 (February 1973): 29–31. Interview documents.

Severing ties with critics, or better yet, punishing financially the institutions that harbor them, is a self-deluding temptation that some officers and defense department officials find hard to resist. Alternative facilities do exist to perform most research tasks. Readily available to the military are government laboratories and defense contractors that are less inclined to tolerate dissenters than are most universities. But a military whose special advantage is technology cannot afford to divorce itself totally from the frontiers of research if it is to be aware of its future opportunities. Moreover, the military cannot expect to maintain the support of society if it is thought to use public money vindictively. In the end, the critics are not silenced, but only strengthened by such a policy.

ONR ABANDONED

With its prime constituency increasingly disinterested in its fate, ONR became vulnerable to raids by bureaucratic rivals. The Bay Saint Louis incident is a good illustration of the organization's stature within the Navy during the mid-1970s.

Bay Saint Louis, Mississippi, was the site of a NASA rocket motor test facility that fell into disuse when the Apollo flights were completed. Both the Army and the Navy sensed that there was advantage in moving employment to the site. The senior senator from Mississippi, John Stennis, was then the chairman of the Senate Armed Services Committee. The Army proposed the transfer of a munitions factory; the Navy offered to relocate its Naval Oceanographic Office and some related activities.

The main work of the Naval Oceanographic Office is the maintenance of information on ocean conditions and the preparation of charts for fleet and contractor use. Previously located in Suitland, Maryland, it would bring over twelve hundred jobs to Mississippi. The admiral in charge of the agency, the Chief Oceanographer of the Navy, saw the relocation as an opportunity to consolidate within his jurisdiction, and at one site, several oceanographic research activities, including three units that were part of ONR. Permission was granted by senior Navy officials to relocate ONR's Long Range Acoustic Propagation Project, the Acoustic Environmental Support Detachment, and the Ocean Science and Technology Division to Bay Saint Louis. With them went over forty jobs and one-tenth of ONR's budget.[12]

[12] Deborah Shapley, "Navy Oceanographic Move Renewal or Disaster for Basic Research," *Science* 188 (20 June 1975): 1189–91.

Especially disappointing to many in ONR was the loss of the Ocean Science and Technology Division which was ONR's link to basic research in oceanography.[13] The rationale for the Navy to invest in basic research in oceanography is more obvious than it is in most scientific disciplines. The unit responsible for supporting oceanography would now become a small part of a very large applied activity that could easily ignore its existence. Worse yet, it would be located in the rural wetlands of Mississippi on a site originally selected by NASA for its assured isolation, one thousand miles away from Washington, D.C., and even further from most of the division's contractors. Although the Secretary of the Navy claimed the consolidation would "revitalize the Navy's Oceanographic Program,"[14] few in ONR thought that it would do the same for the Navy's role in basic research.

Some protests were voiced. The Maryland congressional delegation forced a hearing on the transfer before the Oceanography Subcommittee of the House Merchant Marine and Fisheries Committee where the housing, school, and race relation fears of reassigned civil servants were discussed. Some of the senior scientific personnel indicated they would refuse the transfer. The Chief of Naval Research, Rear Adm. M. D. Van Orden, did not testify, "choosing" instead to take annual leave. Admiral Van Orden retired shortly thereafter, a year ahead of schedule, admitting frustration over his inability to convince senior officials to withdraw the transfer.[15] Only a few of the academic oceanographers bothered to lobby against the decision; 80 percent of their research money came from other government agencies such as NSF and the National Oceanic and Atmospheric Administration in the Department of Commerce.[16]

Once nearly the only expression of governmental interest in basic research, ONR has become a relatively small agency amidst many supporting research. Its reputation for the wise management of research is fading along with a generation of scientists and university administrators. The Navy, when it can be bothered with small sums, will wonder again why ONR exists. Yet, the problem on which ONR always claimed to have been working—the appropriate linkage between science and naval needs—remains to be solved. Neither the Bay Saint Louis approach nor the one academic scientists recall so fondly is likely to provide the answer.

[13] Interview document.

[14] "Navy Announces Consolidation of Oceanographic Program," press release, Department of the Navy, Washington, D.C., 23 July 1975.

[15] "Navy Mississippi Move Approved," Science 189 (15 August 1975): 536.

[16] "A Tale of Two Cities," Science 197 (16 September 1977): 1164.

THE CIVILIZATION OF SCIENCE

The 1980s have been marked by a significant expansion in federal
R&D expenditures, most of which were directed toward increased
weapon development efforts. Overall budget authority for agency-sup-
ported R&D rose from $33 billion in FY 1980 to nearly $61 billion in
FY 1988, an increase of 85 percent in current dollars and 26 percent
in constant dollars. Defense R&D (DOD plus the portion of Department
of Energy allocations devoted to weapons-related activities) grew from
$15 billion to $40 billion in the same period, a 169 percent increase in
current dollars and an 83 percent real increase. In contrast, nondefense R&D declined 24 percent in constant dollars between FY 1980
and FY 1988.[17] The Reagan administration obviously favored military
expenditures over civilian expenditures.

However, the situation in basic research has been quite different.
Defense-supported basic research grew $300 million from FY 1980 to
FY 1988, only an 11 percent real increase over the $600 million base.
Nondefense basic research jumped from $4.2 billion to $8.6 billion, a
40 percent increase measured in constant dollars. Defense now accounts for just one-tenth of federal basic research allocations.[18] While
the administration cut back on civilian agency applied research and
development activities, it encouraged substantial growth in civilian
basic research activities. Following orthodox economics in this instance, the administration argued that the market could provide for
civilian product and process innovation, but not for basic research.[19]

Defense support of university research is still substantial, amounting to $1.3 billion in FY 1988. But it is offered largely for applied re-
search activities conducted most often in special government-spon-
sored laboratories that are loosely affiliated with a few universities.
Basic research allocations for on-campus work total only about $450
million, about the level of the 1960s. ONR remains the largest defense
sponsor of academic science, but it no longer has a dominant role in
the financing of most academic science and engineering fields. Increasingly, it is becoming a contract management agent for other de-

[17] Albert H. Teich, et al., "Federal R&D Funding Since FY 1980," in Intersociety
Working Group, *Research and Development FY 1989* (Washington, D.C.: American Association for the Advancement of Science, 1988), 3–11.

[18] David E. Sanger, "U.S. Research Sponsorship Soars at Universities," *New York
Times*, 8 September 1986, B8; David Graham, "The Pentagon and Basic Research,"
Technology Review 91 (May/June 1988): 12–13.

[19] George A. Keyworth II, "Four Years of Reagan Science Policy: Notable Shifts in
Priorities," *Science* 224 (6 April 1984): 9–13; George A. Keyworth II, "Science and
Technology Policy: The Next Four Years," *Technology Review* 88 (February/March
1985): 45–46, 48, 50–53.

fense agencies, diminishing further its direct influence in academic science (see the appendix).

The major federal sponsors of academic research are NIH and NSF. Alone the NIH provides more than half of the federal funds allocated to basic research at universities. NSF's contribution substantially exceeds that of all defense agencies combined. The Department of Energy, NASA, and the Department of Agriculture have academic research programs that rival those of the Department of Defense. Attempts by the Defense Science Board and similar groups to persuade the military to again increase investments in academic research, have been largely unsuccessful despite the flood of defense dollars that have been available.[20]

THE DEMOCRATIZATION OF SCIENCE

The rationales that garner the most support for science are no longer national security rationales, but rather health and economic growth. Scientists cultivate, or at least do not discourage, expectations that increased allocations for science will lead to significant improvements in health status and economic prosperity. However, the problem for science is that these rationales are much less protective of its independence than are national security rationales in the long run. Whereas politicians are willing to defer to the judgment of the military on most research matters, they have little hesitation to substitute their own for that of civilian officials in nearly all domestic affairs. Recall that Senator Russell did question the likely geographic distribution of NSF awards, but not those of ONR. And recall that it was the internal politics of the Navy that delivered an unhappy Ocean Science and Technology Division to Mississippi, and not a specific request by Senator Stennis for a research agency to populate the vacant Bay Saint Louis facilities.

Some civilian science allocations are naturally pleasing to congressmen. The big NIH budget divides fairly well geographically because most of the money goes to medical schools which are distributed ac-

[20] Report of the Defense Science Board, *Technology Base Management*, Office of the Under Secretary of Defense for Acquisition, August 1987; "Pentagon Opposes Giving Academic Scientists a Larger Share of Defense Research Budget," *The Chronicle of Higher Education*, 13 May 1987, 37; "Enhancing Basic Research," *Aviation Week and Space Technology*, 6 May 1985, 11; "Defense Department Skeptical as House Panel Urges More University Research," *Aviation Week and Space Technology*, 6 May 1985, 101–2. See also Congressional Research Service, *Science Support by the Department of Defense*, a report prepared for the Task Force on Science Policy, Committee on Science and Technology, U.S. House of Representatives, 99th Congress, 2d Session, December 1986, 61–63.

cording to population in order to meet their need for clinical training cases. First-rate medical schools exist in every region of the country, attracting federal research grants with ease. Boston is a major center for medical research, but so too are Atlanta, Houston, Seattle, and Salt Lake City. A multibillion dollar budget assures that no area need feel neglected.

But if geographic equity is not an issue, then the content of research programs can be. At the urging of interest groups, Congress is constantly mandating special attention within NIH for particular diseases and disciplines. The specialists in aging have their own institute, as do the victims of arthritis. Soon there is likely to be an institute for deafness to complement the one for blindness.[21] Sickle-cell anemia, AIDS, and Tay-Sachs disease are only a few of many maladies that are given identifiable research allocations. NIH prides itself in its control through peer review of specific research awards, but it could not resist demands for a major initiative in cancer that went substantially beyond what most cancer specialists thought appropriate.[22] A program in artificial heart research was begun and recently was renewed on the threat of intervention by congressional supporters.[23] The allocations for AIDS research now rival those for cancer and heart disease. NIH continually searches for relief from politics, but never finds it.[24]

The physical sciences suffer similar political pressures. The siting of large-scale research facilities such as particle accelerators and electronics centers become the prize in struggles among regions. The Southwest gets what the Midwest wanted. Scientists have learned to encourage the competition by promising gains in employment for the winning location. The scope of political interest has broadened recently to include university research laboratories and equipment purchases, with Congress enacting legislation requiring agencies to award grants to designated institutions. This earmarking of awards

[21] Julie Rovner, "House Panel OKs Bills on Nursing, Health Issues," *Congressional Quarterly* 46 (2 July 1988): 1837; David Holzman, "Lobbyists' Place in the Laboratory," *Insight*, 24 October 1988, 49–51.

[22] Richard A. Rettig, *The Cancer Crusade: The Story of the National Cancer Act of 1971* (Princeton, N.J.: Princeton University Press, 1977); Mark E. Rushefsky, *Making Cancer Policy* (Albany, N.Y.: State University of New York Press, 1986).

[23] James H. Maxwell, David Blumenthal, and Harvey M. Sapolsky, "Obstacles to Developing and Using Technology: The Case of the Artificial Heart," *International Journal of Technology Assessment in Health Care* 2, no. 3 (1986): 411–24; Harvey M. Sapolsky, "Government and the Development of the Artificial Heart," *The Totally Implantable Artificial Heart*, the report of the Artificial Heart Assessment Panel, National Heart and Lung Institute, National Institutes of Health, June 1973, Appendix A; Barbara J. Culliton, "Politics of the Heart," *Science* 241 (15 July 1988): 283.

[24] National Institute of Medicine, *Responding to Needs and Scientific Opportunity: The Organizational Structure of the National Institutes of Health* (Washington, D.C.: National Academy Press, 1984).

exceeded $330 million in FY 1988.[25] Requests for operating support are certain to follow the acquisition of facilities. And there have been calls to limit the amount that universities in any state could receive in particular programs to satisfy geographic distribution concerns. Massachusetts and California should not hold favored status in the division of research funds, it is argued. Pork barrel politics has discovered science.[26]

Attempts to contain the pressure to spread scientific wealth are undermined both by the desires of particular universities to improve their comparative standing and by the widespread belief that science enhances economic prosperity. Ambitions to gain status among research institutions has led some universities that feel disadvantaged in the normal competition for federal research awards to turn to Congress for direct assistance. Congressmen, eager to aid constituents and convinced by the arguments of scientists that research investments are vital to future economic growth, intervene directly in the award of federal funds, guaranteeing favorable treatment for well-connected, if not yet distinguished, universities. Political review increasingly displaces peer review in research support decisions.[27]

Senior naval officers, like the senior officers in the other armed services, have never been persuaded that investments in science are necessary to assure national security. They have either not believed in the asserted linkage between basic research and new technology or they think it is too long-term a process to be of consequence. ONR's role in postwar science was a role that they did not intend or want; they tolerated its continued existence only when it took on a more military guise, and then quite reluctantly.

In contrast, politicians have come to believe, perhaps too well, in the benefits of basic research investments. Lacking national security justifications for the support of basic research, they find it difficult to contain their enthusiasm. They want their constituents, their causes, to gain a share of the claimed fruits. Scientists, once nurtured solicitously by ONR, are having to adjust to the demands of democratic politics. Science without the Navy, it turns out, is a much less independent, a much more political, undertaking.

[25] "Biggest Pork Barrel Ever," *The Chronicle of Higher Education*, 27 January 1988, 1; "Adapting to Pork-Barrel Science," *Science* 238 (18 December 1987): 1639–40; Phil Kuntz, "Grants for Research Facilities: Experiment to End 'Earmarks,'" *Congressional Quarterly* 46 (2 July 1988): 1824–25.

[26] John Walsh, "Geographic Limit on Research Funds in Bill Seen as Swipe at Peer Review," *Science* 238 (11 December 1987): 1506.

[27] Lois R. Ember, "Efforts to Stem Pork-Barrel Science Funding Likely to be Unsuccessful," *Chemical and Engineering News*, 18 July 1988, 7–16. For an analysis arguing that democratization of support is beneficial for science, see David Eli Drew, *Strengthening Academic Science* (New York: Praeger, 1985).

Budget Data

MONEY may be money, but not in the federal government. To count, federal dollars have to be categorized by agency, appropriation title, fiscal year, and use. Since the Second World War, billions of dollars have flowed through ONR and other agencies in support of academic research. Here I give the dollars their federal identity.

Surprisingly, it is not an easy task to reconstruct federal budget data. Historical budget series are available, but they are not readily made compatible due to changing definitions and the fact that they were assembled for specific purposes, almost always for something other than what I have in mind. Moreover, budgets are subject to political fashion. An agency can inflate a figure for basic research one year and curtail it the next because its officials believe political preferences have shifted. There is flexibility in every budget definition, no matter how elaborate the technical instructions.

Table A–1 describes ONR budgets from FY 1947 through FY 1969 in current dollars. Major external events such as the Korean War and the Sputnik crisis are clearly visible, causing budget spurts that are then eroded away. The Contract Research Program (column one), was the main mechanism for supporting university research. It was supplemented by transfers from other agencies (column four), primarily the Atomic Energy Commission in the early years and then the Department of Defense's Advanced Research Projects Agency in the later years. Support for affiliated laboratories (NRL, the Naval Training Devices Center, etc.) is added to the contract program for ONR R&D (column two). The addition of administrative expenses gives a yearly total for ONR, excluding transfers (column three). Although the budget for ONR's contract research program seems modest by current standards, it exceeded the allocations of NIH for academic medical research until the mid-1950s, and those of NSF for universities until the early 1960s.

With a different format, Table A–2 brings the budget description through FY 1986. The overlap between the tables reveals the extent to which ONR, the sponsor of most of the Navy's basic research, was involved in applied research activities. In 1965, for example, ONR financed about $51 million in basic research, but both its contract research program and the Navy's total support of university research exceeded $83 million. Applied research and development support

TABLE A–1

ONR Funding History, FY 1947–1969
(in millions of dollars)

FY	Contract Research Program[a]	ONR R&D[b]	ONR Total[c]	Contract Research Transfers to ONR from Other Agencies[d]
1947	22.3	43.8	45.9	——
1948	14.2	32.6	34.7	——
1949	19.8	42.3	45.4	10.8
1950	20.5	41.6	44.9	5.7
1951	50.2	77.7	82.2	12.5
1952	35.7	67.2	73.0	16.7
1953	33.7	59.0	65.2	19.1
1954	28.1	53.2	59.3	11.2
1955	30.4	53.7	59.7	12.7
1956	32.2	72.4	79.1	15.5
1957	29.5	85.5	93.0	21.0
1958	51.9	80.2	88.3	19.3
1959	51.5	87.5	95.6	18.5
1960	82.8	118.7	127.3	15.4
1961	70.8	111.9	121.0	16.8
1962	73.2	117.8	127.0	33.6
1963	77.6	127.2	137.3	44.1
1964	78.5	138.9	149.2	42.0
1965	83.1	139.4	151.0	42.7
1966	85.6	148.5	160.4	44.2
1967	82.6	154.5	168.6	not available
1968	89.9	164.5	not available	not available
1969	96.8	168.4	not available	not available

Source: Office of Naval Research.

[a] Contract Research Program primarily involved university research.

[b] Includes support for Naval Research Laboratory and other affiliated laboratories.

[c] Includes ONR administration and construction activities.

[d] Transfers from other agencies for university research.

grew to equal basic research by the mid-1970s and maintained approximately that relationship into the 1980s. It is notable how stagnant the Navy's basic research support was until the mid-1970s, with the beginning of the recent defense buildup and the need to compensate for growing inflation. The end of the buildup is also visible beginning in FY 1986.

The Navy's role as the largest supporter of extramural basic research within the Department of Defense is shown in Table A–3. Only the Air Force begins to rival Navy spending on basic research, but it never gets much beyond two-thirds of the Navy's effort. The Army

Table A–2
Navy Support of University Research, FY 1964–1986
(in millions of dollars)

FY	Basic Research[a,b]	Total University Research Support[c]
1964	49.8	83.6
1965	50.9	83.0
1966	52.0	82.4
1967	52.1	77.6
1968	55.4	88.0
1969	58.3	89.7
1970	48.5	72.5
1971	49.6	70.3
1972	51.9	79.8
1973	45.4	75.9
1974	45.5	57.2
1975	47.9	58.0
1976 & T[d]	64.2	171.3
1977	62.7	151.8
1978	70.8	163.0
1979	86.4	192.8
1980	100.2	207.7
1981	115.0	228.6
1982	142.3	251.3
1983	152.2	272.6
1984	164.6	273.8
1985	176.8	300.7
1986	170.4	276.0

Source: Office of Naval Research.
[a] Fundamental research, 6.1 under DOD definition.
[b] Includes some allocations to nonprofits.
[c] 6.1 through 6.6 under DOD definition.
[d] T is transition quarter due to change in fiscal year.

lags behind the Air Force and the Navy, but with some reason, as it always has a smaller overall budget than the other armed services during peacetime. DARPA's contribution is small, at least in terms of basic research support.

Additional comparative data is shown in Figures A–1 and A–2 and in Table A–4. The 1980s were marked by a significant increase in defense R&D. Figure A–1 outlines the growth in defense R&D as a share of total federal R&D support, with defense rising from about the 50 percent level to nearly the 80 percent level. But when the focus is basic research as a percentage of total federal R&D expenditure, the defense level drops, while that of nondefense increases sharply. Fed-

TABLE A–3

Department of Defense Funding for
University Basic (6.1) Contract Research, FY 1974–1985[a]
(in millions of dollars)

FY	Army	Air Force	Navy	DARPA	Total
1974	13.7	23.2	45.5	21.9	104.3
1975	13.4	22.9	47.0	19.4	103.6
1976	19.0	28.2	64.2	19.1	130.5
1977	23.7	41.0	62.7	18.7	146.1
1978	28.1	49.5	70.8	17.9	166.3
1979	32.0	46.4	86.4	21.0	185.8
1980	38.1	55.5	100.2	19.6	213.4
1981	46.5	63.4	115.0	27.3	252.2
1982	56.1	71.5	142.3	39.4	309.3
1983	71.4	90.3	152.2	46.4	360.3
1984	80.6	112.1	164.6	53.9	411.3
1985(E)[b]	83.8	119.1	176.8	42.7	422.3

Sources: *Defense Appropriations for 1986, Part 8*, Armed Services Committee, U.S. House of Representatives (99th Congress, 2d Session), p. 825; Office of Naval Research.

[a] Restricted to awards exceeding $25,000; grants not included.

[b] Estimate.

eral nondefense R&D support has concentrated on basic research during the 1980s, whereas defense support has concentrated on development activities. Translated back to the total Department of Defense effort, basic research has declined significantly as a share of defense R&D activities.

Table A–4 reports changes between FY 1980 and FY 1988 in terms of current and constant dollars. Defense development activities showed remarkable growth during this period, but so did nondefense basic research support. Nondefense applied research and development support was curtailed as was defense applied research. Defense basic research gained, but even then it accounted for only one-tenth of total federal basic research support in FY 1988. Nondefense basic research now represents about half of all nondefense R&D due to its growth and sharp reductions in nondefense applied research and development support.

The declining role of the Department of Defense in federally-sponsored basic research at universities is described in Table A–5. The department accounted for 44 percent of federal basic research sponsorship in FY 1958. Now, because of the growth of NIH and NSF, it is 11 percent. FY 1975 was the post–World War II low point for defense support of basic research. However, the recent buildup in defense re-

FIGURE A–1. Military R&D as a Percentage of All Federal R&D, 1965–1985.

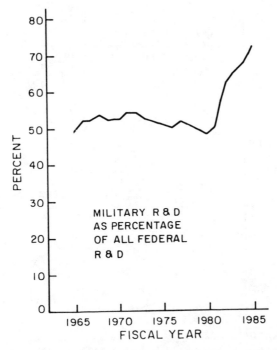

Source: "American Science and Science Policy Issues," Committee on Science and Technology, U.S. House of Representatives (99th Congress, 2d Session), p. 51.

search hardly brought a return to the past dominance of defense sponsorship of university science.

The decline in defense support of applied research at universities was even greater than that of defense support of basic research. As Table A–6 illustrates, Department of Defense support of applied research dropped from $173 million in constant dollars in FY 1965 to $48 million in 1975, and has yet to recover to much more than half of the FY 1965 level. Basic research support is back to the FY 1965 level, but represents a smaller share of total federal sponsorship than it once did, as the other tables demonstrate.

From the late 1940s to the early 1960s the Department of Defense, and especially ONR, was a major source of support for academic research. Pressure from within the department, Congress, and the universities has changed the role of the military in academic research significantly. So too has the growth in civilian agency support and in industry sponsorship of academic research. ONR's money once made a difference in most fields; today, it hardly counts alone or in combination with other defense agencies in any but a few fields of research.

Figure A–2. Basic Research as a Percentage of all Federal R&D, 1965–1985.

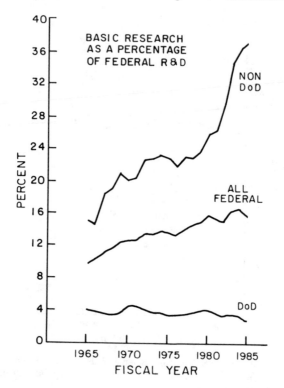

Source: "American Science and Science Policy Issues," Committee on Science and Technology, U.S. House of Representatives (99th Congress, 2d Session), p. 54.

TABLE A–4

Defense and Nondefense R&D by Character of Work
(budget authority in billions of dollars)

	FY 1980 Actual	FY 1988 Estimated	Percent Change Current $	Percent Change Constant $
Defense R&D	$15.0	$40.3	169	83
Basic Research	0.6	0.9	64	11
Applied Research	1.9	2.6	38	−7
Development	12.5	36.7	194	99
Nondefense R&D	$16.7	$18.8	13	−24
Basic Research	4.2	8.6	107	40
Applied Research	5.0	6.5	29	−13
Development	7.5	3.7	−50	−66

Source: AAAS Report VI and "OMB data for Special Analysis J, FY 1989 Budget."
Notes: Includes conduct of R&D only.
Columns may not add up exactly due to rounding.
Percentages based on unrounded numbers.

TABLE A–5

Percentage of Federal Basic Research Obligations
to Colleges and Universities
(in millions of dollars)

FY	Total Federal	DOD	Percent DOD
1958	$ 126	$ 56	44
1960	237	75	32
1965	634	132	21
1970	808	127	16
1975	1,261	106	8
1980	2,320	208	9
1985 estimate	4,022	453	11

Source: Science Support by the Department of Defense, Committee on Science and Technology, U.S. House of Representatives (99th Congress, 2d Session), December 1986, Table 5.2, p. 135.

TABLE A–6

DOD Sponsorship of Basic and Applied Research
at Colleges and Universities
(in millions of dollars)

FY	Basic Research		Applied Research	
	Current $	Constant $[a]	Current $	Constant $
1965	132	176	130	173
1970	127	140	45	49
1975	106	86	59	48
1980	208	117	104	59
1985 estimate	453	193	191	92

Source: Science Support by the Department of Defense, Committee on Science and Technology, U.S. House of Representatives (99th Congress, 2d Session), December 1986, Tables 5.5 and 5.6, pp. 146–47.

[a] Constant 1972 dollars.

Index